W9-ACG-968

FRONTIERS IN
CHEMICAL ENGINEERING

RESEARCH NEEDS
AND OPPORTUNITIES

Committee on Chemical Engineering Frontiers:
Research Needs and Opportunities
Board on Chemical Sciences and Technology
Commission on Physical Sciences, Mathematics, and Resources
National Research Council

NATIONAL ACADEMY PRESS
Washington, D.C.   1988

NATIONAL ACADEMY PRESS ● **2101 Constitution Avenue, NW** ● **Washington, DC 20418**

NOTICE: The project that is the subject of this report was approved by the Governing Board of the National Research Council, whose members are drawn from the councils of the National Academy of Sciences, the National Academy of Engineering, and the Institute of Medicine. The members of the committee responsible for the report were chosen for their special competences and with regard for appropriate balance.

This report has been reviewed by a group other than the authors according to procedures approved by a Report Review Committee consisting of members of the National Academy of Sciences, the National Academy of Engineering, and the Institute of Medicine.

The National Academy of Sciences is a private, nonprofit, self-perpetuating society of distinguished scholars engaged in scientific and engineering research, dedicated to the furtherance of science and technology and to their use for the general welfare. Upon the authority of the charter granted to it by the Congress in 1863, the Academy has a mandate that requires it to advise the federal government on scientific and technical matters. Dr. Frank Press is president of the National Academy of Sciences.

The National Academy of Engineering was established in 1964, under the charter of the National Academy of Sciences, as a parallel organization of outstanding engineers. It is autonomous in its administration and in the selection of its members, sharing with the National Academy of Sciences the responsibility for advising the federal government. The National Academy of Engineering also sponsors engineering programs aimed at meeting national needs, encourages education and research, and recognizes the superior achievements of engineers. Dr. Robert M. White is president of the National Academy of Engineering.

The Institute of Medicine was established in 1970 by the National Academy of Sciences to secure the services of eminent members of appropriate professions in the examination of policy matters pertaining to the health of the public. The Institute acts under the responsibility given to the National Academy of Sciences by its congressional charter to be an adviser to the federal government and, upon its own initiative, to identify issues of medical care, research, and education. Dr. Samuel O. Thier is president of the Institute of Medicine.

The National Research Council was organized by the National Academy of Sciences in 1916 to associate the broad community of science and technology with the Academy's purposes of furthering knowledge and advising the federal government. Functioning in accordance with general policies determined by the Academy, the Council has become the principal operating agency of both the National Academy of Sciences and the National Academy of Engineering in providing services to the government, the public, and the scientific and engineering communities. The Council is administered jointly by both Academies and the Institute of Medicine. Dr. Frank Press and Dr. Robert M. White are chairman and vice-chairman, respectively, of the National Research Council.

Support for this project was provided by the American Chemical Society, the American Institute of Chemical Engineers, the Council for Chemical Research, Inc., the U.S. Department of Energy under Grant No. DE-FG01-85FE60847, the National Bureau of Standards under Grant No. 50SBNB5C23, the National Science Foundation under Grant No. CBT-8419184, and the Whitaker Foundation.

Library of Congress Cataloging-in-Publication Data

National Research Council (U.S.). Committee on Chemical Engineering
  Frontiers: Research Needs and Opportunities.
    Frontiers in chemical engineering : research needs and
  opportunities/Committee on Chemical Engineering Frontiers—
  Research Needs and Opportunities, Board on Chemical Sciences and
  Technology, Commission on Physical Sciences, Mathematics, and
  Resources, National Research Council.
        p.   cm.
    Bibliography: p.
    Includes index.
    ISBN 0-309-03793-X (paper); ISBN 0-309-03836-7 (cloth)
    1. Chemical engineering—Research—United States.   I. Title.
  TP171.N37 1988
  620′ .0072—dc19   88-4120
                                                              CIP
                                                            (Rev.)

Printed in the United States of America

Committee on Chemical Engineering Frontiers: Research Needs and Opportunities

Panels of the Committee

Panel on Biochemical and Biomedical Engineering

ARTHUR E. HUMPHREY (*Chairman*), Lehigh University
KENNETH B. BISCHOFF, University of Delaware
CHARLES BOTTOMLEY, E.I. du Pont de Nemours and Company, Inc.
STUART E. BUILDER, Genentech, Inc.

ROBERT L. DEDRICK, National Institutes of Health
MITCHELL LITT, University of Pennsylvania
ALAN S. MICHAELS, North Carolina State University
FRED PALENSKY, Minnesota Mining and Manufacturing Co., Inc.

Panel on Electronic, Photonic, and Recording Materials and Devices

LARRY F. THOMPSON (*Chairman*), AT&T Bell Laboratories
LEE F. BLYLER, AT&T Bell Laboratories
JAMES ECONOMY, IBM Almaden Research Center
DENNIS W. HESS, University of California, Berkeley

RICHARD POLLARD, University of Houston
T. W. RUSSELL FRASER, University of Delaware
MICHAEL SHEPTAK, Ampex Corporation

Panel on Advanced Materials

ARTHUR B. METZNER (*Chairman*), University of Delaware
FRANK BATES, AT&T Bell Laboratories
C. F. CHANG, Union Carbide Corporation
F. NEIL COGSWELL, Imperial Chemical Industries
WILLIAM W. GRAESSLEY, Princeton University

FRANK KELLEY, University of Akron
JOHN B. WACHTMAN, JR., Rutgers University
IOANNIS V. YANNAS, Massachusetts Institute of Technology

iv

Panel on Energy and Natural Resources Processing

KEITH McHENRY (*Chairman*), Amoco Oil Company

LESLIE BURRIS, Argonne National Laboratory

ELTON J. CAIRNS, Lawrence Berkeley Laboratory

NOEL JARRETT, Alcoa Laboratories

FREDERIC LEDER, Dowell Schlumberger

JOHN SHINN, Chevron Research Company

REUEL SHINNAR, City College of New York

PAUL B. WEISZ, University of Pennsylvania

Panel on Environmental Protection, Safety, and Hazardous Materials

ADEL SAROFIM (*Chairman*), Massachusetts Institute of Technology

SIMON L. GOREN, University of California, Berkeley

GREGORY J. MACRAE, Carnegie Mellon University

THOMAS W. PETERSON, University of Arizona

WILLIAM RODGERS, Oak Ridge National Laboratory

GARY VEURINK, Dow Chemical Company

RAY WITTER, Monsanto Corporation

Panel on Computer Assisted Process and Control Engineering

ARTHUR W. WESTERBERG (*Chairman*), Carnegie Mellon University

HENRY CHIEN, Monsanto Corporation

JAMES M. DOUGLAS, University of Massachusetts

BRUCE A. FINLAYSON, University of Washington

ROLAND KEUNINGS, University of California, Berkeley

MANFRED MORARI, California Institute of Technology

JEFFREY J. SIIROLA, Eastman Kodak Company

WILLIAM SILLIMAN, Exxon Production Research Company

Panel on Surface and Interfacial Engineering

ALEXIS T. BELL (*Chairman*), University of California, Berkeley

RICHARD C. ALKIRE, University of Illinois

JOHN C. BERG, University of Washington

L. LOUIS HEGEDUS, W. R. Grace and Company

ROBERT JANSSON, Monsanto Corporation

KLAVS F. JENSEN, University of Minnesota

JAMES R. KATZER, Mobil Research and Development Company

LEIGH E. NELSON, Minnesota Mining and Manufacturing Company

LANNY D. SCHMIDT, University of Minnesota

Board on Chemical Sciences and Technology

Commission on Physical Sciences, Mathematics, and Resources

Contents

FRONTIERS IN
CHEMICAL ENGINEERING

RESEARCH NEEDS
AND OPPORTUNITIES

ONE

Executive Summary

CHEMICAL ENGINEERING occupies a special place among scientific and engineering disciplines. It is an engineering discipline with deep roots in the world of atoms, molecules, and molecular transformations. The principles and approaches that make up chemical engineering have a long and rich history of contributions to the nation's technological needs. Chemical engineers play a key role in industries as varied as petroleum, food, artificial fibers, petrochemicals, plastics, ceramics, primary metals, glass, and specialty chemicals. All these depend on chemical engineers to tailor manufacturing technology to the requirements of their products and to integrate product design with process design. Chemical engineering was the first engineering profession to recognize the integral relationship between design and manufacture, and this recognition has been one of the major reasons for its success.

This report demonstrates that chemical engineering research will continue to address the technological problems most important to the nation. In the chapters that focus on these problems, many of the discipline's core research areas (e.g., reaction engineering, separations, process design, and control) will appear again and again. The committee hopes that by discussing research frontiers in the context of applications, it will illustrate both the intellectual excitement and the practical importance of chemical engineering.

The research frontiers discussed in this report can be grouped under four overlapping themes: starting new technologies, maintaining leadership in established technologies, protecting and improving the environment, and developing systematic knowledge and generic tools. These frontiers are described in detail in Chapters 3 through 9. From among these, the committee has selected eight high-priority topics that merit the attention of researchers, decision makers in academia and industry, and organizations that fund or otherwise support chemical engineering. These high-priority areas are described below. Recommendations from the committee for initiatives that would permit chemical engineers to exploit these areas are briefly stated in Chapter 10 and detailed in Appendix A.

RESEARCH FRONTIERS IN CHEMICAL ENGINEERING

Starting New Technologies

Chemical engineers have an important role to play in bringing new technologies to commercial fruition. These technologies have their origin in scientific discoveries on the atomic and molecular level. Chemical engineers understand the molecular world and are skilled in integrating product design with process design, process control, and optimization. Their skills are needed to develop genetic engineering (biotechnology) as a manufacturing tool and to create new biomedical devices, and to design new products and manufacturing processes for advanced materials and devices for information storage and handling. In the fierce competition for world markets in these technologies, U.S. leadership in chemical engineering is a strong asset.

Biotechnology and Biomedicine (Chapter 3)

The United States occupies the preeminent scientific position in the "new" biology. If America is to derive the maximum benefit of its investment in basic biological research—whether in the form of better health, improved agriculture, a cleaner environment, or more efficient production of chemicals—it must also assume a preeminent position in biochemical and biomedical engineering. This can be accomplished by carrying out generic research in the following areas:

• Developing chemical engineering models for fundamental biological interactions.
• Studying phenomena at biological surfaces and interfaces that are important in the design of engineered systems.
• Advancing the field of process engineering. Important generic goals for research include the development of separation processes for complex and fragile bioproducts; the design of bioreactors for plant and mammalian tissue culture; and the development of detailed, continuous control of process parameters by rapid, accurate, and noninvasive sensors and instruments.
• Conducting engineering analyses of complex biological systems.

Electronic, Photonic, and Recording Materials and Devices (Chapter 4)

The character of American industry and society has changed dramatically over the past three decades as we have entered the "information age." New information technologies have been made possible by materials and devices whose structure and properties can be controlled with exquisite precision. This control is largely achieved by the use of chemical reactions during manufacturing. Future U.S. leadership in microelectronics, optical information technologies, magnetic data storage, and photovoltaics will depend on staying at the forefront of the chemical technology used in manufacturing processes. Chemical processing will also be a vital part of the likely manufacturing processes for high-temperature superconductors.

At the frontiers of chemical research in this area are a number of important challenges:

• Integrating individual chemical process steps used in the manufacture of electronic, photonic, and recording materials and devices. This is a key to boosting the yield, throughput, and reliability of overall manufacturing processes.
• Refining and applying chemical engineering principles to the design and control of the chemical reactors in which devices are fabricated.
• Pursuing research in separations applicable to the problem of ultrapurification. The materials used in device manufacture must be ultrapure, with levels of some impurities reduced to the parts-per-trillion level;
• Improving the chemical synthesis and processing of polymers and ceramics;
• Developing better processes for deposition and coating of thin films. An integrated circuit, in essence, is a series of electrically connected thin films. Thin films are the key structural feature of recording media and optical fibers, as well.
• Modeling the chemical reactions that are important to manufacturing processes and studying their dynamics.
• Emphasizing process design and control for environmental protection and process safety.

Microstructured Materials (Chapters 5 and 9)

Advanced materials depend on carefully designed structures at the molecular and microscopic levels to achieve specific performance in use. These materials—polymers, ceramics, and composites—are reshaping our society and are contributing to an improved standard of living. The process technology used in manufacturing these materials is crucial—in many instances more important than the composition of the materials themselves. Chemical engineers can make important contributions to materials design and manufacturing by exploring the following research frontiers:

• Understanding how microstructures are formed in materials and learning how to control the processes involved in their formation.

• Combining materials synthesis and materials processing. These areas have traditionally been considered separate research areas. Future advances in materials require a fusion of these topics in research and practice.

• Fabricating and repairing complex materials systems. Mechanical methods currently in use (e.g., riveting of metals) cannot be applied reliably to the composite materials of the future. Chemical methods (e.g., adhesion and molecular self-assembly) will come to the fore.

Maintaining Leadership in Established Technologies

The U.S. chemical processing industries are one of the largest industrial sectors of the U.S. economy. The myriad of industries listed at the beginning of this chapter are pervasive and absolutely essential to society. The U.S. chemical industry is one of the most successful U.S. industries on world markets. At a time of record trade deficits, the chemical industry has maintained both a positive balance of trade and a growing share of world markets (Figure 1.1). The future international competitiveness of these industries should not be taken for granted. Far-sighted management in industry and continued support for basic research from both industry and government are required if this sector of the economy is to continue to contribute to the nation's prosperity.

In a report of this scope and size, it is not possible to spell out the research challenges faced by each part of the chemical processing industries. For example, the committee has reluctantly chosen to pass over food processing, a multibillion-dollar industry where chemical engineering finds a growing variety of applications. The committee has focused its discussion of challenges to the processing industries on energy and natural resources technologies. These

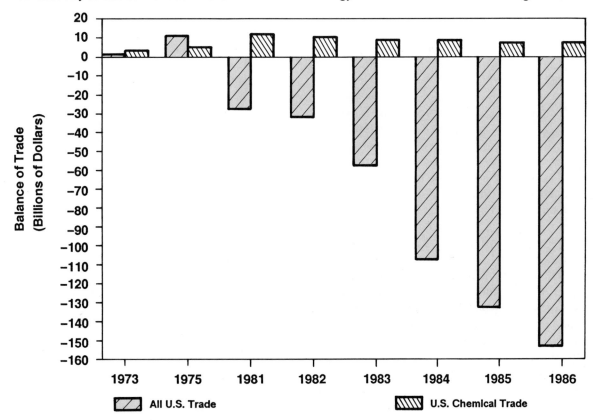

FIGURE 1.1 While the overall U.S. trade balance has plummeted to a deficit of more than $150 billion, the U.S. chemical industry has maintained a positive balance of trade. Courtesy, Department of Commerce.

technologies are key to supplying crucial national needs, keeping the United States competitive, and providing for national security. They are also the focus of substantial research and development in academia and government laboratories, in addition to industry. The committee has identified two high-priority initiatives to sustain the vitality and creativity of engineering research on energy and natural resources. These initiatives focus on in-situ processing of resources and on liquid fuels for the future.

In-Situ Processing of Energy and Mineral Resources (Chapter 6)

The United States has historically benefited from rich domestic resources of minerals and fuels located in readily accessible parts of the earth's crust. These easily reached resources are being rapidly depleted. Our remaining reserves, while considerable, require moving greater and greater amounts of the earth's crust to obtain and process resources, whether that crust is mixed with the desired material (as in a dilute ore vein) or whether it simply lies over the resource. A long-range solution to this problem is to use chemical reactions to extract underground resources, with the earth itself as the reaction vessel. This is known as in-situ processing. Enhanced oil recovery is the most successful current example of in-situ processing, and yet an estimated 300 billion barrels of U.S. oil trapped underground in known reserves cannot be recovered with current technology. Long-range research aimed at oil, shale, tar sands, coal, and mineral resources is needed. Formidable problems exist both for chemists and for chemical engineers. Some research priorities for chemical engineers include separation processes, improved materials, combustion processes, and advanced methods of process design, scale-up, and control. Research on in-situ processing will require collaboration between chemical engineers and scientists and engineers skilled in areas such as geology, geophysics, hydrology, environmental science, mechanical engineering, physics, mineralogy,

materials science, metallurgy, surface and colloid science, and chemistry.

Liquid Fuels for the Future (Chapters 6 and 9)

Our current and foreseeable transportation technologies depend completely on a plentiful supply of liquid fossil fuels. Anticipatory research to ensure a future supply of these fuels is a wise investment. Research of this type subsumes a number of generic challenges in chemical engineering, including:

• Finding new chemical process pathways that can make large advances in the production of liquid fuels from solid and gaseous resources.
• Processing solids, since equipment design and scale-up are greatly limited by our lack of fundamental understanding of solids behavior.
• Developing better separation processes.
• Conducting research on materials capable of withstanding the extreme processing conditions that may be encountered when processing liquid fuels.
• Advancing the state of the art in the design, scale-up, and control of processes.

Protecting and Improving the Environment

Responsible Management of Hazardous Substances (Chapter 7)

The modern world faces many environmental problems. Some of these are a consequence of producing the ever-increasing number and variety of chemicals and materials demanded by society. Chemical engineers must take up the role of cradle-to-grave guardians for chemicals, ensuring their safe and environmentally sound manufacture, use, and disposal. This means becoming involved in a range of research areas dealing with environmental protection, process safety, and hazardous waste management. In the following four areas, the challenges are clear, the opportunities for chemical engineering research are abundant, and the potential benefits to society are great.

• Conducting long-term research on the generation, control, movement, fate, detection, and environmental and health effects of contaminants in the air, water, and land. Chemical engineering research should include the fundamental investigation of combustion processes, the application of biotechnology to waste degradation, the development of sensors and measurement techniques, and participation in interdisciplinary studies of the environment's capacity to assimilate the broad range of chemicals that are hazardous to humans and ecosystems.

• Developing new chemical engineering design tools to deal with the multiple objectives of minimum cost; process resilience to changes in inputs; minimization of toxic intermediates and products; and safe response to upset conditions, start-up, and shutdown.

• Directing research at cost-effective management of hazardous waste, as well as improved technologies (e.g., combustion) or new technologies for destroying hazardous waste.

• Carrying out research to facilitate multimedia, multispecies approaches to waste management. Acid rain and the leaching of hazardous chemicals from landfills demonstrate the mobility of chemicals from one medium (e.g., air, water, or soil) to another.

Developing Systematic Knowledge and Generic Tools

The success of chemical engineers in contributing to a diverse set of technologies is due to an emphasis on discovering and developing basic principles that transcend individual technologies. If, 20 years from now, chemical engineers are to have the same opportunities for contributing to important societal problems that they have today, then the research areas described in the preceding sections must be explored and supported in a way that maximizes the development of basic knowledge and tools.

In surveying the field of chemical engineering, the committee has identified two cross-cutting areas that are in a state of rapid development and that promise major contributions to a wide range of technologies. Accordingly, this report singles out for special attention the advances

under way in applying modern computational methods and process control to chemical engineering and the promise of basic research in surface and interfacial engineering.

Advanced Computational Methods and Process Control (Chapter 8)

The speed and capability of the modern computer are revolutionizing the practice of chemical engineering. Advances in speed and memory size and improvements in complex problem-solving ability are more than doubling the effective speed of the computer each year. This unrelenting pace of advance has reached the stage where it profoundly alters the way in which chemical engineers can conceptualize problems and approach solutions. For example:

• It is now realistic to imagine mathematical models of fundamental phenomena beginning to replace laboratory and field experiments. Such computations increasingly allow chemical engineers to bypass the long (2 to 3 years), costly step of producing process and product prototypes, and permit the design of products and processes that better utilize scarce resources, are significantly less polluting, and are much safer.

• Future computer aids will allow design and control engineers to examine many more alternatives much more thoroughly and thus produce better solutions to problems within the known technology.

• Better modeling will allow the design of processes that are easier and safer to operate. Improved control methodology and sensors will overcome the current inability to model certain processes accurately.

• Sensors of the future will be incredibly small and capable. Many will feature a chemical laboratory and a computer on a chip. They will enable chemical engineers to detect chemical compositions inside hostile process environments and revolutionize their ability to control processes.

To realize the promise of the computer in chemical engineering, we need a much larger effort to develop methodologies for process design and control. In addition, state-of-the-art computational facilities and equipment must become more widely disseminated into chemical engineering departments in order to integrate methodological advances into the mainstream of research and education.

Surface and Interfacial Engineering (Chapter 9)

Surfaces, interfaces, and microstructures play an important role in many of the above-mentioned research frontiers. Chemical engineers explore structure-property relationships at the atomic and molecular level, investigate elementary chemical and physical transformations occurring at phase boundaries, apply modern theoretical methods for predicting chemical dynamics at surfaces, and integrate this knowledge into models that can be used in process design and evaluation. Fundamental advances in these areas will have a broad impact on many technologies. Examples include laying down thin films for microelectronic circuits, developing high-strength concrete for roadways and buildings, and inventing new membranes for artificial organs. Advances in surface and interfacial engineering can also move the field of heterogeneous catalysis forward significantly. New knowledge can help chemical engineers play a much bigger role in the synthesis and modification of novel catalysts with enhanced capabilities. This activity would complement their traditional strength in analytical reaction engineering of catalysts.

HIGHLIGHTS OF THE RECOMMENDATIONS

Education and Training of Chemical Engineers (Chapter 10)

The new research frontiers in chemical engineering, some of which represent new applications for the discipline, have important implications for education. A continued emphasis is needed on basic principles that cut across many applications, but a new way of teaching those principles is also needed. Students must be exposed to both traditional and novel applications of chemical engineering. The American

Institute of Chemical Engineers (AIChE) has set in motion a project to incorporate into undergraduate chemical engineering courses examples and problems from emerging applications of the discipline. The committee applauds this work, as well as recent AIChE moves to allow more flexibility for students in accredited departments to take science electives.

A second important need in the curriculum is for a far greater emphasis on design and control for process safety, waste minimization, and minimal adverse environmental impact. These themes need to be woven into the curriculum wherever possible. The AIChE Center for Chemical Process Safety is attempting to provide curricular material in this area, but a larger effort than this project is needed. Several large chemical companies have significant expertise in this area. Closer interaction between academic researchers and educators and industry is required to disseminate this expertise.

The Future Size and Composition of Academic Departments (Chapter 10)

A bold step by universities is needed if their chemical engineering departments are (1) to help the United States achieve the preeminent position of leadership in new technologies and (2) to keep the highly successful U.S. chemical processing industries at the forefront of world markets for established technologies. The universities should conduct a one-time expansion of their chemical engineering departments over the next 5 years, exercising a preference for new faculty capable of research at interdisciplinary frontiers.

This expansion will require a major commitment of resources on the part of universities, government, and industry. How can such a preferential commitment to one discipline be justified, particularly at a time of budgetary austerity? One answer is that the worldwide contest for dominance in biotechnology, advanced materials technologies, and advanced information devices is in full swing, and the United States cannot afford to stand by until it gets its budgetary house in order. As the uniquely "molecular" engineers, chemical engineers have powerful tools that need to be refined and applied to the commercialization of these technologies. A second answer is that the alternative to expansion, a redistribution of existing resources for chemical engineering research, would cut into vital programs that support U.S. competitiveness in established chemical technologies. The recommendation for an expansion in chemical engineering departments is not a call for "more of the same." It is the most practical way to move chemical engineering aggressively into the new areas represented by this report's research priorities while maintaining the discipline's current strength and excellence.

Balanced Portfolios (Chapter 10)

The net result of an additional investment of resources in chemical engineering should be the creation of three balanced portfolios: one of priority research areas, one of sources of funding for research, one of mechanisms by which that funding can be provided.

The eight priority research areas described above constitute the committee's recommendation of a balanced portfolio of research areas on the frontiers of the discipline.

In terms of a balanced portfolio of funding sources, the committee proposes initiatives for industry and a number of federal agencies in Chapter 10 and Appendix A to ensure a healthy diversity of sponsors. Table 1.1 links specific research frontiers to funding initiatives for potential sponsors.

A third balanced portfolio, of funding mechanisms, is needed if the above-mentioned research frontiers are to be pursued in the most effective manner. Different frontiers will require different mixes of mechanisms, and the decision to use a particular mechanism should be determined by the nature of the research problem, by instrumentation and facilities requirements, and by the perceived need for trained personnel in particular areas for industry. This topic is discussed in more detail in Chapter 10.

The Need for Expanded Support of Research in Chemistry (Chapter 10)

Chemical engineering builds on research results from other disciplines, as well as those from its own practitioners. Not surprisingly, the most important of these other disciplines is chemistry. A vital base of chemical science is needed to stimulate future progress in chemical engineering, just as a vital base in chemical engineering is needed to capitalize on advances in chemistry. The committee endorses the recommendations contained in the NRC's 1985 report *Opportunities in Chemistry*, and urges their implementation in addition to the recommendations contained in this volume.

TABLE 1.1 High-Priority Research Frontiers and Initiatives[a]

Priority Research Area	Relevant Audiences[b]							
	NSF	DOE	NIH	DOD	EPA	NBS	BOM	CPI
Biotechnology and biomedicine	✔✔	✔	✔✔			✔		✔
Electronic, photonic, and recording materials and devices	✔✔	✔		✔✔				✔
Microstructured materials	✔✔	✔✔		✔✔		✔		✔
In-situ processing of resources	✔	✔✔					✔✔	
Liquid fuels for the future	✔	✔✔				✔		✔✔
Responsible management of hazardous substances	✔				✔✔			✔✔
Advanced computational methods and process control	✔✔	✔						✔✔
Surface and interfacial engineering	✔✔	✔✔						✔

[a] ✔✔ = major initiative recommended; ✔ = supporting initiative recommended.

[b] NSF = National Science Foundation; DOE = Department of Energy; NIH = National Institutes of Health; DOD = Department of Defense; EPA = Environmental Protection Agency; NBS = National Bureau of Standards; BOM = Bureau of Mines; CPI = U.S. chemical processing industries.

TWO

What Is Chemical Engineering?

Chemical engineering has a rich past and a bright future. In barely a century, its practitioners have erected the technological infrastructure of much of modern society. Without their contributions, industries as diverse as petroleum processing, pharmaceutical manufacturing, food processing, textiles, and chemical manufacturing would not exist as we know them today. In the 10 to 15 years ahead, chemical engineering will evolve to address challenges that span a wide range of intellectual disciplines and physical scales (from the molecular scale to the planetary scale). And chemical engineers, with their strong ties to the molecular sciences, will be the "interfacial researchers" bridging science and engineering in the multidisciplinary environments where a host of new technologies will be brought into being.

MAGINE A WORLD where penicillin and other antibiotics are rarer and more expensive than the finest diamonds. Imagine countries gripped by famine as dwindling supplies of natural guano and saltpeter cause fertilizers to become increasingly scarce. Imagine hospitals and clinics where kidney dialysis is as risky and as uncertain over the long term as today's artificial heart program. Imagine serving on a police force or in the infantry without a lightweight bulletproof vest. Imagine your closet with no wash-and-dry, wrinkle-free synthetic garments, or your home without durable, easy-cleaning, mothproof synthetic rugs. Imagine cities choked with smog and soot from millions of residential coal furnaces and millions of automobiles without emission controls. Imagine an "information society" trying to function on vacuum tubes and ferrite core storage for data processing. Imagine paying $25 or more for a gallon of gasoline, if you can even buy it. This world, in which few of us would want to live, is what a world without chemical engineering would be like.

Chemical engineers have made so many important contributions to society that it is hard to visualize modern life without the large-volume production of antibiotics, fertilizers and agricultural chemicals, special polymers for biomedical devices, high-strength polymer composites, and synthetic fibers and fabrics. How would our industries function without environmental control technologies; without processes to make semiconductors, magnetic disks and tapes, and optical information storage devices; without modern petroleum processing? All these technologies require the ability to produce specially designed chemicals—and the materials based on them—economically and with a minimal adverse impact on the environment. Developing this ability and implementing it on a practical scale is what chemical engineering is all about.

The products that depend on chemical engineering emerge from a diverse array of industries that play a key role in our economy (Table 2.1). These industries produce most of the materials from which consumer products are made, as well as the basic commodities on which our way of life is built. In 1986, they shipped products valued at nearly $585 billion. They had a payroll of 3.3 million employees, or

TABLE 2.1   The Chemical Processing Industries in the United States[a]

Industry[b]	Number of Employees (thousands)	Value of Shipments ($ millions)	Value Added by Manufacture ($ millions)
Food and beverages	378	73,633	24,370
Textiles	99	7,649	2,897
Paper	322	51,145	19,871
Chemicals	1,023	197,932	95,258
Petroleum	169	129,365	17,112
Rubber and plastics	340	31,078	15,390
Stone, clay, and glass	354	34,372	17,449
Nonferrous metals	49	21,920	524
Other nondurables	577	37,594	24,291
TOTAL	3,321	584,689	217,161
Chemical processing industries' share of total manufacturing	17.5%	25.7%	21.7%

[a] Data for employment and value of shipments are for 1986. Data for value added by manufacture are for 1985. SOURCE: Data Resources, Inc.

[b] The definition of the chemical processing industries (CPI) used in this table is the one used by Data Resources and *Chemical Engineering* in compiling their statistics on these industries. For several of the industries listed, only a part is considered to be in the CPI and data are presented for this part only. A list of the Standard Industrial Classification codes used to define the CPI for this table is given in Appendix C.

17.5 percent of all U.S. manufacturing employees. They generated over $217 billion in value added in 1985, or 21.7 percent of all U.S. manufacturing value added. The chemicals portion of the CPI is one of the most successful U.S. industries in world competition, producing an export surplus of $7.8 billion in 1986, in contrast to the overall U.S. trade deficit of $152 billion.

But chemical engineering is more than a group of basic industries or a pile of economic statistics. As an intellectual discipline, it is deeply involved in both basic and applied research. Chemical engineers bring a unique set of tools and methods to the study and solution of some of society's most pressing problems.

TRADITIONAL PARADIGMS OF CHEMICAL ENGINEERING

Every scientific discipline has its characteristic set of problems and systematic methods for obtaining their solution—that is, its paradigm. Chemical engineering is no exception. Since its birth in the last century, its fundamental intellectual model has undergone a series of dramatic changes.

When the Massachusetts Institute of Technology (MIT) started a chemical engineering program in 1888 as an option in its chemistry department, the curriculum largely described industrial operations and was organized by specific products. The lack of a paradigm soon became apparent. A better intellectual foundation was required because knowledge from one chemical industry was often different in detail from knowledge from other industries, just as the chemistry of sulfuric acid is very different from that of lubricating oil.

Unit Operations

The first paradigm for the discipline was based on the unifying concept of "unit operations" proposed by Arthur D. Little in 1915. It evolved in response to the need for economic large-scale manufacture of commodity products. The concept of unit operations held that any chemical manufacturing process could be resolved into a coordinated series of operations such as pulverizing, drying, roasting, crystallizing, filtering, evaporating, distilling, electrolyzing, and so on. Thus, for example, the academic study of the specific aspects of turpentine manufacture could be replaced by the generic study of distillation, a process common to many other industries. A quantitative form of the unit operations concept emerged around 1920, just in time for the nation's first gasoline crisis. The rapidly growing number of automobiles was severely straining the production capacity for naturally occurring gasoline. The ability of chemical engineers to quantitatively characterize unit operations such as distillation allowed for the rational design of the first modern oil refineries. The first boom of employment of chemical engineers in the oil industry was on.

During this period of intensive development of unit operations, other classical tools of chemical engineering analysis were introduced or were extensively developed. These included studies of the material and energy balance of processes and fundamental thermodynamic studies of multicomponent systems.

Chemical engineers played a key role in helping the United States and its allies win World War II. They developed routes to synthetic rubber to replace the sources of natural rubber that were lost to the Japanese early in the war. They provided the uranium-235 needed to build the atomic bomb, scaling up the manufacturing process in one step from the laboratory to the largest industrial plant that had ever been built. And they were instrumental in perfecting the manufacture of penicillin, which saved the lives of potentially hundreds of thousands of wounded soldiers. An in-depth look at this latter contribution shows the sophistication that chemical engineering had achieved by the 1940s.[1]

Penicillin was discovered before the war, but could only be prepared in highly dilute, impure, and unstable solutions. Up to 1943, when chemical engineers first became involved with the project, industrial manufacturers used a batch purification process that destroyed or inactivated about two-thirds of the penicillin produced. Within 7 months of their involvement, chemical engineers at an oil company (Shell Development Company) had applied their knowledge of generic engineering principles to build a fully integrated pilot plant that processed 200 gallons of fermentation broth per day and achieved nearly 85 percent recovery of penicillin. When this process was installed by four penicillin producers, production soared from a rate in 1943 capable of sustaining the treatment of 4,100 patients per month to a rate in the second half of 1944 equivalent to treatments for nearly 250,000 patients per month.

A second challenge in getting penicillin to the front was its instability in solution. A stable form was needed for storage and shipment to hospitals and clinics. Freeze drying—in which the penicillin solution was frozen to ice and then subjected to a vacuum to remove the ice as water vapor—seemed to be the best solution, but it had never been implemented on a production scale before. A crash project by chemical engineers at MIT during 1942–1943 established enough understanding of the underlying phenomena to allow workable production plants to be built.

The Engineering Science Movement

The high noon of American dominance in chemical manufacturing after World War II saw the gradual exhaustion of research problems in conventional unit operations. This led to the rise of a second paradigm for chemical engineering, pioneered by the engineering science movement. Dissatisfied with empirical descriptions of process equipment performance, chemical engineers began to reexamine unit operations from a more fundamental point of view. The phenomena that take place in unit operations were resolved into sets of molecular events. Quantitative mechanistic models for these events were developed and used to analyze existing equipment, as well as to design new process equipment. Mathematical models of processes and reactors were developed and applied to capital-intensive U.S. industries such as commodity petrochemicals.

THE CONTEMPORARY TRAINING OF CHEMICAL ENGINEERS

Parallel to the growth of the engineering science movement was the evolution of the core

chemical engineering curriculum in its present form. Perhaps more than any other development, the core curriculum is responsible for the confidence with which chemical engineers integrate knowledge from many disciplines in the solution of complex problems.

The core curriculum provides a background in some of the basic sciences, including mathematics, physics, and chemistry. This background is needed to undertake a rigorous study of the topics central to chemical engineering, including:

- multicomponent thermodynamics and kinetics,
- transport phenomena,
- unit operations,
- reaction engineering,
- process design and control, and
- plant design and systems engineering.

This training has enabled chemical engineers to become leading contributors to a number of interdisciplinary areas, including catalysis, colloid science and technology, combustion, electrochemical engineering, and polymer science and technology.

A NEW PARADIGM FOR CHEMICAL ENGINEERING

Over the next few years, a confluence of intellectual advances, technological challenges, and economic driving forces will shape a new model of what chemical engineering is and what chemical engineers do (Table 2.2).

A major force behind this evolution will be the explosion of new products and materials that will enter the market during the next two decades. Whether from the biotechnology industry, the electronics industry, or the high-performance materials industry, these products will be critically dependent on structure and design at the molecular level for their usefulness. They will require manufacturing processes that can precisely control their chemical composition and structure. These demands will create new opportunities for chemical engineers, both in product design and in process innovation.

A second force that will contribute to a new chemical engineering paradigm is the increased

TABLE 2.2 Enduring and Emerging Characteristics of Chemical Engineering

Enduring Characteristics	Emerging Characteristics
Serves industries whose products remain unchanged on the market for long periods	Serves industries whose products are quickly superseded in the marketplace by improved ones
Serves industries that compete mainly on the basis of price and availability	Serves industries that compete on the basis of quality and product performance
Expertise in the manufacture of homogeneous materials from small molecules	Expertise in the manufacture of composite and structured materials from large molecules
Expertise in the manufacture of commodity materials	Expertise in the manufacture of high-performance and specialty materials
Expertise in process design	Expertise in designing products with special performance characteristics
Expertise in designing large-volume processes	Expertise in designing small-scale processes
Expertise in designing continuous processes	Expertise in designing batch processes
Expertise in designing industrial plants dedicated to a single product or process	Expertise in designing flexible manufacturing plants
Practitioners use simple models and approximations to solve problems	Practitioners use more complete models, better approximations, and large computers to solve problems rigorously
Practitioners have access to only a few simple analytical instruments	Practitioners have access to many sophisticated analytical instruments
Practitioners build their careers around a single product line or process	Practitioners have multiple career changes
Academic research is mostly performed by single principal investigators within chemical engineering departments	Academic research is also performed by multidisciplinary groups of principal investigators, sometimes in centers or other organizational environments
Research and education focus on the mesoscale (equipment level)	Research and education also include the microscale (molecular level) and macroscale (systems level)

competition for worldwide markets. Product quality and performance are becoming more important to global competition than ever before. If the United States is to remain competitive in world chemical markets, it must find new ways to lower costs and improve product quality and consistency. Similarly, a strong domestic energy-producing industry is needed to preclude foreign domination of this vital sector of the economy. The key to meeting these challenges is innovation in process design, control, and manufacturing operations. It is particularly important that the United States maintain a vigorous presence in commodity chemical markets. Commodities are at the base of industries that employ millions of Americans, provide basic necessities for our society, and generate valuable export earnings. Thriving commodity businesses are also vital to specialty chemical businesses. The technical expertise and financial resources that commodities provide is crucial to the long-term research and development efforts that specialties require.

The third force shaping the future of chemical engineering is society's increasing awareness of health risks and environmental impacts from the manufacture, transportation, use, and ultimate disposal of chemicals. This will be an important source of new challenges to chemical engineers. Modern society will not tolerate a continuing occurrence of such incidents as the release of methyl isocyanate at Bhopal (in 1985) and the contamination of the Rhine (in 1986). It is up to the chemical engineering profession to act as the cradle-to-grave guardian for chemicals, ensuring their safe and environmentally sound use.

The fourth and most important force in the development of tomorrow's chemical engineering is the intellectual curiosity of chemical engineers themselves. As they extend the limits of past concepts and ideas, chemical engineering researchers are creating new knowledge and tools that will profoundly influence the training and practice of the next generation of chemical engineers.

When a discipline adopts a new paradigm, exciting things happen, and the current era is probably one of the most challenging and potentially rewarding times to be entering chemical engineering. How can the unfolding pattern of change in the discipline be described?

The focus of chemical engineering has always been industrial processes that change the physical state or chemical composition of materials. Chemical engineers engage in the synthesis, design, testing, scale-up, operation, control, and optimization of these processes. The traditional level of size and complexity at which they have worked on these problems might be termed the *mesoscale*. Examples of this scale include reactors and equipment for single processes (unit operations) and combinations of unit operations in manufacturing plants. Future research at the mesoscale will be increasingly supplemented by studies of phenomena taking place at molecular dimensions—the microscale—and the dimensions of extremely complex systems—the macroscale (see Table 2.3).

Chemical engineers of the future will be integrating a wider range of scales than any other branch of engineering. For example, some may work to relate the macroscale of the environment to the mesoscale of combustion systems and the microscale of molecular reactions and transport (see Chapter 7). Others may work to relate the macroscale performance of a composite aircraft to the mesoscale chemical reactor in which the wing was formed, the design of the reactor perhaps having been influenced by

studies of the microscale dynamics of complex liquids (see Chapter 5).

Thus, future chemical engineers will conceive and rigorously solve problems on a continuum of scales ranging from microscale to macroscale. They will bring new tools and insights to research and practice from other disciplines: molecular biology, chemistry, solid-state physics, materials science, and electrical engineering. And they will make increasing use of computers, artificial intelligence, and expert systems in problem solving, in product and process design, and in manufacturing.

Two important developments will be part of this unfolding picture of the discipline:

• Chemical engineers will become more heavily involved in product design as a complement to process design. As the properties of a product in performance become increasingly linked to the way in which it is processed, the traditional distinction between product and process design will become blurred. There will be a special design challenge in established and emerging industries that produce proprietary, differentiated products tailored to exacting performance specifications. These products are characterized by the need for rapid innovation, as they are quickly superseded in the marketplace by newer products.

• Chemical engineers will be frequent participants in multidisciplinary research efforts. Chemical engineering has a long history of fruitful interdisciplinary research with the chem-

TABLE 2.3 Microscale-Mesoscale-Macroscale: Illustrations

Microscale ($\leq 10^{-3}$ m)
    Atomic and molecular studies of catalysts
    Chemical processing in the manufacture of integrated circuits
    Studies of the dynamics of suspensions and microstructured fluids

Mesoscale (10^{-3}–10^2 m)
    Improving the rate and capacity of separations equipment
    Design of injection molding equipment to produce car bumpers made from polymers
    Designing feedback control systems for bioreactors

Macroscale (>10 m)
    Operability analysis and control system synthesis for an entire chemical plant
    Mathematical modeling of transport and chemical reactions of combustion-generated air pollutants
    Manipulating a petroleum reservoir during enhanced oil recovery through remote sensing of process data, development and use of dynamic models of underground interactions, and selective injection of chemicals to improve efficiency of recovery

ical sciences, particularly in industry. The position of chemical engineering as the engineering discipline with the strongest tie to the molecular sciences is an asset, since such sciences as chemistry, molecular biology, biomedicine, and solid-state physics are providing the seeds for tomorrow's technologies. Chemical engineering has a bright future as the "interfacial discipline" that will bridge science and engineering in the multidisciplinary environments where these new technologies will be brought into being.

Some things, though, will not change. The underlying philosophy of how to train chemical engineers—emphasizing basic principles that are relatively immune to changes in field of application—must remain constant if chemical engineers of the future are to master the broad spectrum of problems that they will encounter. At the same time, the way in which this philosophy finds concrete expression in course offerings and requirements must be responsive to changing needs and situations.

NOTE

1. Additional background and references on chemical engineers and the effort to win World War II may be found in *Separation and Purification: Critical Needs and Opportunities* (Washington, D.C.: National Academy Press, 1987), pp. 92–100.

THREE

Biotechnology and Biomedicine

Advances in molecular biology and medicine are spawning new technologies and new opportunities for chemical engineers. Potential areas for contributions to human health include the design and manufacture of artificial organs, diagnostic tests, and therapeutic drugs. In agriculture, the manufacture of veterinary pharmaceuticals and the scaling up of plant cell-culture techniques represent new applications for chemical engineering principles. Other opportunities include using genetically engineered systems for the synthesis of chemicals and the biological treatment of waste. This rich potpourri of technological possibilities has attracted intense interest on the part of the United States' technological competitors. They are putting in place substantial research programs and facilities to exploit the potential of biotechnology. This chapter describes intellectual frontiers that chemical engineers should pursue. They include modeling of fundamental biological interactions, investigating surface and interfacial phenomena important to engineering design in living systems, expanding the scope of process engineering into biological systems, and conducting engineering analysis of whole-organ or whole-body systems. Implications of these new challenges for chemical engineering research and education are discussed.

MERICA LEADS the world in the biosciences, thanks largely to 25 years of major support for fundamental research by the federal government. This research in the "new" biology—aspects of which are popularly known as biotechnology—is providing the basis for revolutions in health care, agriculture, food processing, environmental improvement, and natural resource utilization. The new technologies that will be made possible by advances in the biosciences, and particularly in molecular biology, will be applied to the search for solutions to some of the world's most pressing problems. They will, in addition, create new industries and spur economic growth. Estimates of the potential annual market for new products from these technologies range from $56 billion to $69 billion for the year 2000 (Table 3.1).

CHALLENGES TO CHEMICAL ENGINEERS

The commercialization of developments in biotechnology will require a new breed of chem-

ical engineer, one with a solid foundation in the life sciences as well as in process engineering principles. This engineer will be able to bring innovative and economic solutions to problems in health care delivery and in the large-scale implementation of advances in molecular biology.

The biologically oriented chemical engineer will focus on areas ranging from molecular and cellular biological systems (biochemical engineering) to organ and whole-body systems and processes (biomedical engineering). Biochemical engineers will focus on the engineering problems of adapting the "new" biology to the commercial production of therapeutic, diagnostic, and food products. Biomedical engineers will apply the tools of chemical engineering modeling and analysis to study the function and response of organs and body systems; to elucidate the transport of substances in the body; and to design artificial organs, artificial tissues, and prostheses. These exciting opportunities for chemical engineers are described in more detail below, first in terms of the potential impact on

TABLE 3.1 Estimated World Markets for the Products of Biotechnology (millions of dollars)[a]

Market	Year 1985	1990	2000
Medical products			
Pharmaceuticals		3,500	20,000–30,000
Diagnostics	100	1,500	5,000
Veterinary products	100	1,500	5,000
Other (materials, sensors, etc.)		75	500
Chemicals			
Fine and specialty chemicals		500	2,000–4,000
Commodity chemicals			1,000
Agricultural products			
Chemicals and biologicals	20	500	2,000
Plants and seeds	25	1,500	5,000–6,000
Improved animal breeds	20	500	1,000
Food and animal feed products			
Additives and supplements	200	1,500	4,000
Flavors and fragrances	10	100	500
Associated equipment and engineering systems	1,500	4,000	10,000
TOTAL	1,975	15,175	56,000–69,000

[a] Dollar values are at manufacturer's level. Inflation is estimated at 6 to 8 percent per year.

SOURCE: SRI International.

Improving Platelet Storage

Trauma, leukemia, and hemophilia patients commonly require infusions of platelets to control bleeding. These platelets are obtained from separation from donated whole blood and are stored in special plastic storage bags. Using current storage methods, these platelets survive only 3 days. This results in a chronic shortage of platelets, particularly after weekends or holidays when donations decline. A program that has applied chemical engineering principles to this problem has demonstrated a new material for storage bags that can keep platelets viable for more than a week.

The chemical engineering approach began with an analysis of the biochemistry of platelet metabolism. Like many cells, platelets consume glucose by two pathways, an oxidative pathway and an anaerobic pathway. The oxidative pathway produces carbon dioxide, which makes the solution containing the platelets more acidic (lower pH) and promotes anaerobic metabolism. This second metabolic pathway produces large amounts of lactic acid, further lowering pH. The drop in pH from both pathways kills the platelets.

The chemical engineering solution was to design a new material for the storage bag that was capable of "breathing"—of allowing carbon dioxide to diffuse out and oxygen to diffuse in. This prevents the drop in pH. Platelets stored in this new bag survive 10 days or more.

society and then in terms of intellectual frontiers for research.

Human Health

Chemical engineers are needed to help transform the results of basic health research into practical products. They have been instrumental in designing processes for the safe and economical production of extremely complex therapeutic and diagnostic agents (e.g., insulin and hepatitis-B surface antigen). The insert boxes in this chapter on platelet storage (p. 19), tissue plasminogen activator (p. 21), interferons (p. 29), and kidney function (p. 32) illustrate the significance of chemical engineering research in this area.

Artificial Organs, Artificial Tissues, and Prostheses

Chemical engineers can also make an important contribution to the development of artificial organs, artificial tissues, and prostheses. In fact, the first successful artificial organ—the artificial kidney—was the result of an innovative NIH program in the early 1960s that brought together an interdisciplinary team of chemical engineers, materials scientists, and physicians. Chemical engineers applied the fundamental concepts of fluid mechanics, membrane transport theory, mass transfer, and interfacial physical chemistry to systems design. They developed predictive correlations relating the blood-clearance performance of a dialyzer to operating parameters such as membrane area, channel dimensions, blood and dialysate flow rates, pressure drop in the system, and temperature. Within 5 years, several soundly engineered prototype systems, using disposable membrane cartridges and sophisticated monitoring and control equipment, were in advanced stages of development. By the mid-1970s, hemodialysis had graduated from an experimental procedure to a well-established, reliable, and safe means of sustaining patients suffering from acute and chronic renal failure. Today, hemodialysis and its companion process, hemofiltration, are standard hospital and clinical procedures and are responsible for major reductions in mortality and morbidity due to kidney failure (Plate 1).

The success of the artificial kidney can be attributed to the relative simplicity of its task. Unwanted substances are removed through a membrane separation carried out in a device external to the body. Some of the targets for future artificial organs, such as the pancreas and the liver, are much more complex systems in which significant numbers of chemical reactions are carried out. In these cases, replacement might take the form of hybrid artificial organs containing living and functional cells in an artificial matrix. Development of such systems will be critically dependent on the contributions of chemical engineers to interdisciplinary teams.

The concept of the artificial pancreas illustrates how chemical engineers can develop new artificial or semiartificial organs, particularly if they are grounded in whole-organ physiology and biochemistry and capable of communicating fluently with endocrinologists and physiologists. A chemical engineer working alone might conceive of an implantable power-driven insulin pump, for instance, controlled by feedback from an electronic glucose sensor. In talking with an endocrinologist, the engineer might devise an implantable device containing viable pancreatic islet cells that functions as a normal pancreas. Working with a subcellular physiologist and enzymologist, the chemical engineer might come up with what is, in effect, an artificial islet cell— a "smart membrane" device that senses blood glucose levels and in response releases insulin from a reservoir encapsulated by the membrane. Each of these design concepts is potentially useful; the one that ultimately succeeds in practice will be the one that is easiest to make, most reliable, and most durable under the actual conditions of use. The wide choice of options and alternatives makes this field of research particularly exciting and rewarding for chemical engineers.

Artificial organs that perform the physical and biochemical functions of the heart, liver, pancreas, or lung are one class of organ replacements. A rather different target of opportunity is the development of biological materials that play a more passive role in the body; for example,

• biocompatible polymer solutions whose rheological properties make them suitable as replacements for synovial fluids in joints or the aqueous and vitreous humors in the eye;
• temporary systems that stimulate the regeneration of lost or diseased body mass and then are absorbed or degraded by the body (e.g., an artificial "second skin" for burn patients); and
• electrochemical signal transduction systems that would allow the body's nervous system to communicate with and control musculoskeletal prostheses.

Diagnostics

A second area rich in opportunities for chemical engineers is the design and manufacture of diagnostic systems and devices. Molecular biologists have discovered or created a variety of enzymes and monoclonal antibodies that are capable of detecting a wide range of diseases, disorders, and genetic defects. Chemical engineers are needed to incorporate these materials into devices and systems that are fast, inexpensive, accurate, and not susceptible to error on the part of the person carrying out the test. For example, although an enzyme-linked immunosorbant assay (ELISA) exists for detecting the antibodies to cytomegalovirus (CMV) in blood samples, it cannot be reliably used in practice to follow the course of a new CMV infection. The error introduced into the test by having different operators perform it on each new blood sample in the series is sufficient to render highly questionable the interpretation of trends in the series, particularly if changes in the magnitude of the result are small. It is important to be able to follow trends in CMV antibodies because CMV infections can be life-threatening to individuals with compromised immune systems, and congenital CMV infections are the single largest cause of birth defects.

Chemical engineering research leading to the design of devices and systems that are fast and "accurate" includes the following:

• development of selectively adsorbent, functionalized porous media to which immunoreagents can be affixed and that are amenable to speedy optical assay after contact with body fluids;
• design of fluid-containing substrates that allow small volumes of test fluids to contact reagents efficiently and with highly reproducible assay response; and
• design of flexible manufacturing systems to make the wide variety of expensive monoclonal antibodies needed for diagnostic test kits.

Chemical engineers at several pharmaceutical firms are using hollow fiber reactors to grow monoclonal antibody-producing hybridomas in an in vitro batch process. Research on reactor design to optimize the production of monoclonal antibodies will have a significant impact on the future development, economy, and use of diagnostic tests.

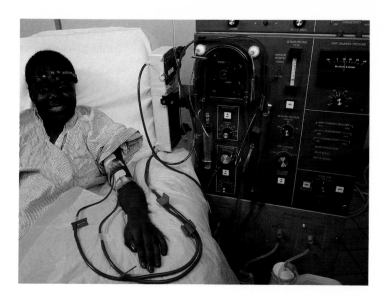

PLATE 1 The kidney dialysis machine (artificial kidney) is responsible for major reductions in deaths and adverse health consequences from kidney failure. Its development required a team effort that brought together chemical engineers, physicians, and materials scientists. The design of the disposable filter cartridge, shown attached to the front left side of the dialysis machine, was a major contribution by chemical engineers to the project. Courtesy, National Institute of Diabetes and Digestive and Kidney Diseases.

PLATE 2 These crystals of human insulin are made by bacteria whose genetic instructions have been altered using recombinant DNA techniques. Human insulin is needed by diabetics who develop allergies to the animal-derived insulin that has been used to treat the disease since 1921. Without chemical engineering contributions such as process design and purification technology, though, the large-scale production of human insulin would not be possible. Courtesy, Eli Lilly and Company.

PLATE 3 Cell culture of plants now takes place in individual containers cared for by hand. Automation of plant cell culture using chemical engineering techniques could improve the yields and economics of plant cell culture and expand the range of its applications in the production of new species and hybrids. Courtesy, Monsanto Company.

PLATE 4 Bioreactors that use mammalian cells, like this tower fermentor, are on the cutting edge of new biotechnology manufacturing processes. Courtesy, Cetus Corporation.

Tissue Plasminogen Activator: Superior Therapy For Heart Attacks

Many serious health problems result from abnormally located blood clots: heart attacks (clots in coronary arteries), pulmonary embolism (clots in the lungs), and peripheral arterial occlusion and deep vein thrombosis (clots in the limbs). Each year heart attacks alone afflict over a million people in the United States, and almost half of them die as a result.

In the past, only two treatments have been available for breaking down blood clots: streptokinase and urokinase. Both treatments lack specificity for clots, so they can cause a general breakdown of the hemostatic system, sometimes leading to generalized bleeding.

Recently, a superior therapy has been approved for use by the federal government: tissue-type plasminogen activator (tPA). This naturally occurring enzyme dissolves blood clots as part of the normal healing process. By administering relatively large quantities of it, clot breakdown time can be shortened from about a week to under an hour.

Normal circulating levels of tPA are low, so that to accomplish this dramatic clot breakdown one would need the amount of tPA contained in 50,000 liters of blood. This is clearly not practical. Instead, the molecule has been cloned and expressed in mammalian cells so that it can be produced in quantity. Using cells from mammals, rather than bacteria, results in a product molecule that has the same folding, internal bonding, and coat of sugar residues as the natural protein.

Producing the kilograms of tPA necessary to satisfy the world's therapeutic needs requires the special skills possessed by modern biochemical engineers. Sophisticated engineering of the fermentation vessels, culturing conditions, and media compositions is required to culture thousands of liters of mammalian cells. In addition, new extremes of purity must be achieved in order to assure the safety of proteins derived from mammalian cells. The cost of the starting materials and the capacity constraints of the present-day equipment require that yields from each fermentation batch be as high as possible.

The current cost per dose of tPA (about $1,000) has already emerged as an important barrier to its widespread use in hospitals and clinics. Continued research in chemical engineering will be crucial to finding more economical processes for the production of this breakthrough therapeutic.

Preventing and Curing Disease

The biological activity of many of the next generation of compounds needed to prevent disease (e.g., vaccines) or to cure it (e.g., drugs) will depend on precisely designed three-dimensional configurations. These configurations can be most easily created by synthesizing the compounds biologically or from biologically derived precursors, using cells that have been altered through recombinant DNA techniques (Plate 2). The manufacture of these compounds, examples of which are listed in Table 3.2, will entail new challenges for chemical engineers. For processes involving bacteria or yeast as product sources, the manufacture of molecules with the correct three-dimensional configuration may require additional steps to modify or refold the proteins. Processes involving plant and mammalian tissue cultures as product sources will require new types of reactors capable of growing the specialized cells, control procedures and sensors tailored for biological processing, and extremely special and gentle purification procedures to ensure that products of adequate purity can be produced without chemical change or loss of configuration. These are formidable engineering problems. Chemical engineers, long involved in the manufacture of antibiotics, peptides, and simple proteins, have significant experience to apply to these problems.

Providing new modes of delivering drugs presents almost as important an opportunity as providing new ways of making them. The standard practice of periodically administering drug doses can lead to initial concentrations in the body that may be sufficiently high to induce undesirable side effects. Later, as the drug is metabolized or eliminated, its concentration can drop below the effective level (Figure 3.1). This

TABLE 3.2 Important Therapeutic Targets of Opportunity

Therapeutic	Action
Antigens	Stimulate antibody response
Interferons	Regulate cellular response to viral infections and cancer proliferation
Tissue plasminogen activators	Stop thrombosis by dissolving blood clots
Human growth hormone	Reverse hypopituitarism in children
Neuroactive peptides	Mimic the body's pain-controlling peptides
Regulatory peptides	Stimulate regrowth of bone and cartilage
Lymphokines	Modulate immune reactions
Human serum albumin	Treat physical trauma
Gamma globulin	Prevent infections
Antihemophilic factors	Treat hereditary bleeding disorders
Monoclonal antibodies	Provide site-specific diagnostics and drug delivery

problem is particularly important with drugs that are metabolized or eliminated rapidly from the body and with drugs that have a narrow therapeutic range (the span between the therapeutically effective and the toxic concentrations). The optimal pharmacological effect can sometimes be attained by establishing and maintaining a steady-state concentration of the drug or by time-sequencing its administration. The controlled release of short-half-life drugs over a long period of time can be effected by administering the drug through low-flow pumps, as a mixture of capsules that disintegrate at different rates, or in pouches inserted under the eyelid or taped to the skin (Figure 3.2). Chemical engineers have been instrumental in designing and manufacturing polymers that are capable of such controlled release over long periods of time.

Another approach to delivering drugs is to target the administration of a drug to a specific site in the body. This might be accomplished by coupling a drug to an antibody that has been cloned to attack a specific receptor at the disease site. This approach would make possible, for example, the selective exposure of tumor-bearing tissues to high concentrations of toxic drugs. Chemical engineers are needed to produce such targeted drugs and to elucidate the kinetics of monoclonal antibody transport through the body to target sites.

Other areas in therapeutics that are ripe for interdisciplinary collaboration include the design of special-purpose pumps and catheters, sterile implants that allow access from outside the body to veins and body organs, and imaging techniques for monitoring drug levels. Efforts by chemical engineers to provide improved data acquisition and quantitative modeling of pharmacokinetics can lead to the design of better drug administration procedures and better timing to maximize the delivery of drugs to the organs that need them while minimizing the exposure of other organs.

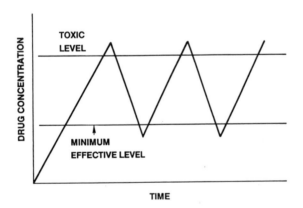

FIGURE 3.1 When a tablet of medicine is taken, or an injection given, sharp fluctuations of drug levels in the body can result. At the peak level, undesirable side effects of the drug can manifest themselves. Unless the tablet or injection is given very frequently, the level of the drug in the body can fall too low to be effective. Chemical engineers are working on ways to deliver drugs that maintain a steady, effective level of the drug in the body.

Agriculture

Major opportunities exist for chemical engineers to help develop agricultural biochemicals.

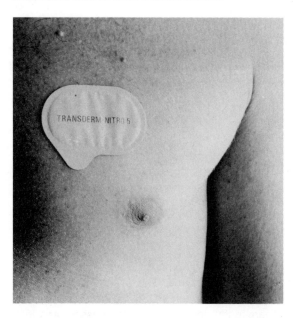

FIGURE 3.2 This transdermal (through the skin) product delivers a steady level of nitroglycerin to the body, preventing the pain of angina. The thin, adhesive unit administers the drug directly to the bloodstream when applied to the skin. This once-a-day patch provides medication without interfering in a patient's daily activities or without having to take pills several times a day. It does not require puncturing the skin with a needle. Chemical engineers are involved in the design and manufacture of new polymer systems for medical applications such as this. Courtesy, ALZA Corporation.

These opportunities roughly parallel the frontiers that have opened up in the human health area. In agriculture, a deeper understanding of biological processes in plants has paved the way for biologically derived fungicides and herbicides that are highly potent, species specific, and environmentally safe. The rapid introduction of these compounds into widespread use will require expertise in process design, process control, and separation technology to ensure that they are manufactured free from contaminants that would threaten the environment or worker safety.

A second focus for chemical engineers in agriculture is the improvement of veterinary pharmaceuticals (e.g., peptide hormones that promise to stimulate growth, fecundity, and feed efficiency in farm animals) and vaccines. The prospects for improvement of these compounds parallel the bright prospects for human pharmaceuticals and vaccines, and the requirements for chemical engineering expertise are similar.

A third focus is the development of large-scale plant-cell culture techniques. These techniques convert undifferentiated cell clumps into differentiated cells of genetically selected roots and stems ready for planting. Such plant cell clones are already being used to produce new crop varieties that are more resistant to adverse environmental conditions or disease. Examples include disease-resistant trees and virus-free potatoes. Cell culture techniques will continue to be used to increase crop productivity by allowing horticulturists to propagate quickly new plant strains showing

- increased resistance to pests, drought, or soil salinity;
- higher productivity or enhanced growth rates;
- ability to produce increased amounts or higher quality of seed proteins and other plant products such as alkaloids, carotenes, latex, and steroids; and
- improved efficiency of nitrogen fixation and photosynthesis.

At present, cell culture work is done mostly by hand by horticulturists in large greenhouses (Plate 3). Chemical engineers could greatly increase the usefulness of this method of plant propagation by developing efficient automated processes for producing plants from cloned cells.

Biochemical Synthesis

By manipulating the genetic machinery of the cell, it is possible to cause most cellular systems to produce virtually any biochemical material. Unfortunately, the growth of cellular systems (particularly in tissue cultures) is constrained by end-product inhibition and repression; hence, it is difficult to produce end products in high concentration. Furthermore, cells are always grown in aqueous solution, so biochemicals produced by cellular routes must have intrinsically high value in order for the cost of recovery from dilute aqueous solution to be minimized. Thus, most biochemicals of commercial interest to be produced by biotechnology will be high-value products such as enzymes, biopolymers, or metabolic cofactors. In general, their potency is so high that only small quantities will be needed. Accordingly, the challenge to chemical engineers in producing these products is not so much in process scale-up but rather in obtaining high process yield and minimal process losses.

Enzymes are an important class of biochemicals; they are the catalysts needed in the chemical reaction cycles of living systems, and they execute their catalytic role with exquisite chemical precision. Enzymes have great potential in synthetic chemistry because they can effect stereospecific reactions, avoiding the production of an unwanted isomer of a complex molecule. Currently, many of the enzymes used in industrial processing (e.g., those used to convert starch into sugar or milk into cheese) are derived from microbial sources because they are beyond the practical reach of current synthetic chemical technology. Biotechnology offers the potential, through cellular genetic control, for making enzymes—not only

---

High-Fructose Corn Syrup: Biotechnology on a Billion-Dollar Scale

When you crack open a can of Coca Cola or Pepsi, you are tasting some of the fruits of biochemical engineering! Most non-diet soft drinks sold in the United States are sweetened with high-fructose corn syrup (HFCS), a substitute for the natural sugar that comes from cane and beets. HFCS, produced by an enzymatic reaction, is an example of the successful application of chemical engineering principles to biochemical synthesis. So successful, in fact, that more than $1.5 billion of HFCS was sold in the United States last year.

To make HFCS a commercial reality, two separate bioprocesses had to be developed, scaled up, and brought on line in a manufacturing plant. The first bioprocess was a fermentation to manufacture the necessary enzyme. The second process used the enzyme to convert dextrose to HFCS. The early involvement of chemical engineers in the design of these processes, and their fruitful interaction with biologists, was a key to the success of these two endeavors.

The fermentation for making isomerase enzyme is relatively fast and can be carried out in a number of process configurations. Basic to all of the process configurations are the problems of maintaining sterility and containment; engineering heat and mass transfer; controlling levels of O_2 and CO_2 in the fermentation solution; and regulating temperature, pressure, and levels of dextrose. The solution must be carefully mixed during the fermentation; damage to the cells by agitation can either mechanically kill the microorganisms that produce the enzyme or hopelessly complicate the recovery of the enzyme from the fermentation broth. Chemical engineers are skilled in solving all of these problems.

The conversion of dextrose syrups to high-fructose corn syrups

---

those that are now used industrially but also others for new uses in synthetic chemistry. The synthesis and processing of these complex molecules require conditions that will maintain their specific three-dimensional structures. One challenge for chemical engineers will be to develop processes that can meet the rigorous requirements for optimally producing and recovering enzymes.

Another challenge will be to understand the chemical transformations that enzymes catalyze. The goal would be to determine how these transformations can be used or tailored through changes in enzyme structure to produce compounds that are difficult or costly to produce

by traditional synthetic chemistry. Addressing this challenge will bring the chemical engineer into close contact with biochemists and synthetic chemists.

Environment and Natural Resources

Biotechnology offers promise for improving the quality of our environment through the introduction of new microbial and enzymatic techniques for removing and destroying toxic pollutants in municipal and industrial wastes. This opportunity is discussed in detail in Chapter 7.

The depletion of domestic high-grade ore deposits has made the United States vulnerable requires two additional chemical engineering steps. The first is the rigorous purification of the dextrose to remove any contaminants that could inactivate the isomerase enzyme. The dextrose syrup is rigorously demineralized, filtered, and refined over carbon, treated with a magnesium co-catalyst, and brought to the appropriate temperature and level of acidity. The dextrose is then ready for the second step. It is passed down a column containing the isomerase enzyme isolated on a carrier. Enzyme loadings of over 10 million units of active enzyme per cubic foot are not uncommon. The isomerization process can be conducted in either of two ways, depending on market conditions. In the summer, when the demand is great, it is common to run dextrose through the columns as quickly as possible to supply the customer. This method of operation results in higher column temperatures that shorten the life of the enzyme. The columns need to be changed more frequently, but the market demand can be met without building excess manufacturing capacity. In the winter, when the demand is low, the manufacturer reduces the flow to get maximum enzymatic lifetime and, correspondingly, lowest operating costs. Enzymatic lifetime can be detrimentally affected by poor control of temperature and acidity, and air or any insoluble material that can plug the column will shorten the lifetime of the column regardless of enzyme activity. The economics of HFCS production, thus, are based on the ability to maintain superb process control.

The record time in which HFCS was developed and brought to high levels of production and sales is a testament to the versatility and power of chemical engineering principles. No new chemical engineering principles had to be discovered to make HFCS a commercial reality. They were waiting to be applied to a biological system.

to shortages of metals (e.g., chromium, manganese, and niobium) that are important to the production of high-strength steel and other alloys. Biological systems with a high affinity for metals are known, and genetically engineered microorganisms might be used to sequester metals from highly dilute waste streams (see Chapter 6), from dilute sources underground (see Chapter 6), or from the sea. To make such recovery concepts practical, chemical engineering will be needed to design systems that allow these microorganisms to function optimally and to efficiently contact large volumes of dilute solutions, or, in the case of in-situ metals extraction, to operate efficiently

when the earth itself is the bioreactor.

Another opportunity for biotechnology may be to provide a new source for certain petrochemicals. Biological routes to a number of organic chemicals currently derived from petroleum have been demonstrated (Table 3.3). For structurally complex chemicals, these routes may prove more economically efficient than alternative routes (e.g., those using synthesis gas from coal gasification as a starting material). Whether this will be the case depends largely on engineering research efforts in bioprocessing and in other resource areas.

INTERNATIONAL COMPETITION

Who will lead the commercialization of the "new" biology? The answer is not yet clear. Our principal technological competitors in the world, the Europeans and the Japanese, are aggressively expanding their efforts to commercialize the results of basic biological research. West Germany, Japan, and the United Kingdom each have three large government-supported institutes dedicated to biotechnology. The United States has only one center of comparable magnitude (the MIT Biotechnology Process Center). Not surprisingly, our competitors are establishing commercial positions by practicing effective and forward-looking biochemical and biomedical engineering. Some examples of their recent accomplishments attest to their aggressiveness.

• Basic technology for membrane separation of biomolecules was invented in the United States, but the West Germans and the Japanese lead in its application to separations of enzymes and amino acids from complex mixtures. Jap-

TABLE 3.3 Potential Routes to Commodity Chemicals by Microbial Fermentation of Glucose

Chemical	Microorganism(s)
Ethanol	*Saccharomyces cerevisiae*
	Zymomonas mobilis
Butanol	*Clostridium acetobutylicum*
Adipic acid	*Pseudomonas* species
Methyl ethyl ketone	*Klebsiella pneumoniae*
Glycerol	*Saccharomyces cerevisiae*
	Dunaliella species
Citric acid	*Aspergillus niger*

SOURCE: T. K. Ng, R. M. Busche, C. C. McDonald, and R. W. F. Hardy, *Science*, 219, 1983, 733. Copyright 1983 by the AAAS. Excerpted with permission.

anese government support of membrane separation research and development alone amounted to $21 million in 1983. This is many times the level of comparable effort expended by the U.S. government. One impact of the well-funded Japanese effort can be seen in the increasing number of Japanese kidney dialyzers appearing in U.S. hospitals.

• Technology for very large (400,000-gallon) continuous fermenters was developed and is being practiced in the United Kingdom. This development pushes biochemical engineering to limits not yet explored in the United States.

• Although the use of fermentation to produce ethanol is an ancient technology, more efficient immobilized-cell, continuous processes have been conceived, and Japan has established the first demonstration-scale plant.

According to the Office of Technology Assessment (OTA), Western Europe and Japan have historically maintained a large and stable funding pattern for biochemical engineering. This is not so for the United States. The existing base of biochemical engineers in other countries, and their strong interest in exploiting the discoveries of the "new" biology, are reflected by extensive government funding and facilities support. It is clear that countries such as West Germany and Japan are laying a foundation of engineering research and training as part of their overall strategy for intense international com-

petition in biotechnology and medicine. The potential economic rewards for success are very great, as shown in Table 3.1. First entry into these markets will be critically important in international competition, and major shares in the worldwide bioproducts market will be captured by those countries who possess the needed research infrastructure.

INTELLECTUAL FRONTIERS

The intellectual frontiers for chemical engineers in biotechnology and biomedicine can be described on a continuum from microscale through mesoscale to macroscale. At either end of this spectrum are highly interdisciplinary research topics that will require modeling and analytical tools currently used by chemical engineers in other contexts. The important mesoscale challenges of bioprocessing will require chemical engineering expertise in reaction engineering, process design and control, and separations. The following sections discuss these challenges in greater detail.

Models for Fundamental Biological Interactions

The living microbial, animal, or plant cell can be viewed as a chemical plant of microscopic size. It can extract raw materials from its environment and use them to replicate itself as well as to synthesize myriad valuable products that can be stored in the cell or excreted. This microscopic chemical plant contains its own power station, which operates with admirably high efficiency. It also contains its own sophisticated control system, which maintains appropriate balances of mass and energy fluxes through the links of its internal reaction network.

Cell membranes are not simply passive containers for the cell's contents. Rather, they are highly organized, dynamic, and structurally complex biological systems that regulate the transfer of specific chemicals through the cell wall.

One important constituent of cell membranes is a class of molecules—the phospholipids—that spontaneously form two-layer films in a number of geometries. Many of the important physical properties of cell membranes, such as two-dimensional diffusion and differentiation between the inside and the outside of a tube or sphere, can be studied with these spontaneously formed structures.

If we can develop accurate quantitative models that simulate how cells respond to various environmental changes, we can better utilize the chemical synthesis capabilities of cells. Steps toward this goal are being taken. Models of the common gut bacterium *Escherichia coli* have been developed from mechanisms of subcellular processes discovered or postulated by molecular biologists. These models have progressed to the point where they can be used with experiments to discriminate among postulated mechanisms for control of subcellular processes.

Some of the most promising potential applications of biotechnology involve animal or plant cells. Models for these organisms, which have greater internal complexity as well as more demanding environmental requirements than simple cells, are not yet available. It will probably be necessary to incorporate the structure of functional subunits of the cell (organelles) into models for complex cells in addition to the chemical structure that is used in bacterial cell models. Cellular reactions are subject to the limitations imposed by the laws of thermodynamics, by diffusion, and by reaction kinetics. Chemical engineers are familiar with the techniques for designing mathematical models that involve these parameters and should be able to make major contributions to the development of cellular models. The development of reliable models hinges on acquiring accurate data bases on enzymes, biologically important proteins, and cellular systems. The data should include physical properties, transport properties, chemical properties, and reaction rate information.

Biological Surfaces and Interfaces

Many biological reactions and processes occur at phase boundaries and are thus controlled by surface interactions. Examples include such highly efficient processes as selective transport of ions across membranes, antibody-antigen interactions, cellular protein synthesis, and nerve impulse transmission. Progress in achieving similar efficiencies in engineered enzyme processes, bioseparations, and information transmission can be aided by acquiring more sophisticated knowledge of biochemical processes at interfaces. With this knowledge, such products as synthetic antibodies for human and animal antigens, or synthetic membranes that can serve as artificial red blood cells or transport barriers, could be developed.

Surface interactions play an important role in the ability of certain animal cells to grow and produce the desired bioproducts. An understanding of the dynamics of cell surface interactions in these "anchorage-dependent" cells (cells that function well only when attached to a surface) will be needed, for example, to improve the design of bioreactors for growing animal cells.

Interactions at surfaces and interfaces also play an essential role in the design and function of clinical implants and biomedical devices. With a few recent exceptions, implants do not attach well to tissue, and the resulting mobility of the tissue-implant interface encourages chronic inflammation. The result can be a gathering of platelets at the site, leading to a blood clot or to the formation of a fibrous capsule, or scar, around the implant (Figure 3.3).

A number of fundamental questions about biological changes at the tissue-implant interface challenge chemical engineers in the design of medical implants and devices. How do cells interact with the surfaces of well-characterized materials? Which receptor sites on cell membranes interact with which functional groups on the surfaces of biomedical materials? What is the effect of other morphological features of the surface, or of the mechanical properties of the material? How does the metabolic activity of the cell change after a reaction with a material interface? What new enzymes or chemicals are produced by the cell after such a reaction? How does information gained in this area lead to better materials, or to the development of new methods for attaching biomedical materials to tissues? How can chemical engineers contribute to better ways of monitoring implanted materials noninvasively?

Bioprocessing

Three major intellectual frontiers for chemical engineers in bioprocessing are the design of bioreactors for the culture of plant and animal cells, the development of control systems along with the needed biosensors and analytical instruments, and the development of processes for separating and purifying products. A critical component in each of these three research areas is the need to relate the microscale to the mesoscale.

Bioreactors for Manufacturing Processes

Much of the early work in applying recombinant DNA technology to the production of bioactive substances has used microbial cell species such as bacteria, yeasts, and molds. These microbes are fairly easy to manipulate genetically and are hardy under adverse conditions. Unfortunately, animal or plant proteins produced by clones of microbial cells often lack the critical three-dimensional structure that is formed when the same proteins are produced by animal and plant cells. For this reason, these proteins may not be biologically active even though they have the correct sequence of amino acids. One important future area of biotechnology lies in using plant and animal cells in place of microbial cells. The large-scale use of plant and animal cells in tissue culture raises important problems in the design and operation of bioreactors (Plate 4).

One problem mentioned earlier is that certain animal cells are anchorage-dependent. Also, plant and animal cells are easily ruptured by mechanical shear. Bioreactors for handling such cells must be designed so that the contents of the reactor can be mixed without disrupting the cells. A similar problem exists in the design of

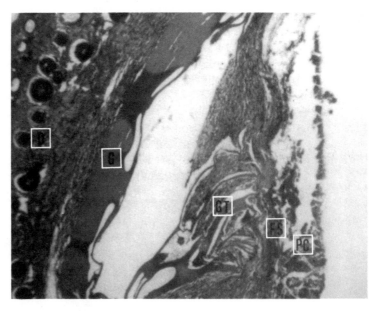

FIGURE 3.3 Implanted materials and devices in the body that are perceived as foreign objects will encourage the formation of scar tissue surrounding them. In this photomicrograph, a nonporous membrane (G), implanted between the skin and some subcutaneous tissue (D and PC), generates the synthesis of granulation tissue (GT) and a fibrotic sac (scar tissue, FS) within 4 weeks. Courtesy, Ioannis Yannas, Massachusetts Institute of Technology.

systems to transfer the cells from one vessel to another.

Plant cells tend to aggregate, and large aggregates pose problems in maintaining a supply of nutrients to all cells and in removing wastes. The development of bioreactors for plant cells will require an understanding of limitations on mass transfer in such aggregates.

Some bioreactor systems must be completely protected from microbial contamination, meaning that not a single alien bacterium or virus particle can be allowed to penetrate the system. Reliable and economical systems need to be developed to achieve this level of contamination prevention. Along with the need for prevention is the need to be able to detect contamination at a level of a few microorganisms in a hundred kiloliters of medium. This degree of detection is not yet achievable. Research could vastly improve the crude detection methods that are used today.

Most industrial bioprocesses are now operated in a batch mode. Batch processing is the method of choice for small-scale production,

Interferons

Cancer now afflicts one out of four adults. One of the more promising therapies for certain kinds of cancers involves the use of interferon, a protein that occurs in minute quantities in the body where it is an essential part of the body's immune system. Interferon can be produced outside the body in cultures of transformed lymphoblastoid cells. A few years ago, it was possible to culture these human cells on scales up to a few hundred milliliters. Chemical engineers have now developed reactors for the aseptic culture of human cells on a scale 100,000 times larger, making it possible to produce human interferon in practical useful quantities.

The problems encountered in this scale-up illustrate the difficulties of engineering a biological process. It is possible to achieve a 20 percent yield of interferon from human cell culture. Yet, the first attempts at large-scale production gave yields of less than 5 percent. This was because a by-product of the cell culture was a protease—an enzyme that breaks down proteins such as interferon. Chemical engineers cooperated with biochemists in designing protease inhibitors and made a crucial contribution to the problem by designing a continuous, short-run process that minimized exposure of the interferon to the protease. Chemical engineers have also developed chromatographic separation systems that improve the yield by separating interferon from the reaction mixture quickly and efficiently.

As medical researchers continue to explore the therapeutic properties of interferons, chemical engineers will continue to provide the expertise needed to make available the quantities of these molecules necessary for clinical evaluation.

and it has the advantage that the equipment can be used for intermittent production of more than one product. An intriguing future possibility is that chemicals and biochemicals will be produced by biotechnology on a large-scale, continuous basis. Continuous processing frequently offers advantages in economy and uniformity of product quality. However, the engineering problems involved in converting from batch to continuous biological processing are not trivial. Continuous processing of biological systems places stringent demands on equipment design, instrumentation, and operation for maintaining aseptic conditions and biological containment. One indication of these difficulties is the fact that although processes for fermenting natural materials to produce beer predate written history, beer is still brewed and aged in batches. Attempts to use a continuous process to manufacture a product as well understood as beer have not produced a beverage with acceptable taste.

Process Monitoring and Control

Continuous and detailed knowledge of process conditions is necessary for the control and optimization of bioprocessing operations. Because of containment and contamination problems, this knowledge must often be obtained without sampling the process stream. At present, conditions such as temperature, pressure, and acidity (pH) can be measured rapidly and accurately. It is more difficult to monitor the concentrations of the chemical species in the reaction medium, to say nothing of monitoring the cell density and intracellular concentrations of hundreds of compounds.

The development of rapid, accurate, and noninvasive on-line measurement sensors and instruments is a high-priority goal in the commercialization of biotechnology (Figure 3.4). Some of these instruments will build on analytical methods now used in catalysis and other surface sciences, such as

- Fourier transform infrared spectroscopy,
- fluorospectrometry,
- mass spectrometry,
- nuclear magnetic resonance (NMR) spectrometry, and
- combinations of some of the above-mentioned methods with chromatography.

These methods will be applied by chemical engineers to monitor and control reaction and recovery systems.

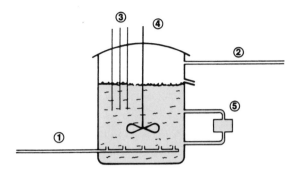

FIGURE 3.4 Several approaches to developing analytical instrumentation for bioreactors are shown in this figure. (1) Gases being fed to the bioreactor must be analyzed to determine their flow rate and composition. Flow rates can be measured with mass flowmeters or rotameters; the concentrations of oxygen and carbon dioxide in the gas mixture can be determined by electrochemical methods or IR analysis, respectively. These data, when combined with similar measurements on the gases exiting from the bioreactor (2), provide information on oxygen uptake and carbon dioxide evolution in the bioreactor. (3) Various sensors may be placed in the bioreactor. Properties that might be measured include temperature, pressure, pH, dissolved oxygen, and liquid feed rates. Sensors are under development to measure glucose, ethanol, various ions (e.g., NH_4^+, Mg^{2+}, K^+, Na^+, Cu^{2+}, and PO_4^{3-}), and other important biomolecules (e.g., ATP, ADP, AMP, DNA, RNA, and NADH). (4) The rotating shaft of the impeller can be used to measure viscosity. (5) Varous spectrophotometric cells can be used to measure properties such as turbidity, if the culture medium in the bioreactor is not too dense. From L. E. Erickson and G. Stephanopoulos, "Biological Reactors," ch. 12 in *Chemical Reaction and Reactor Engineering*, J. J. Carberry and A. Varma (eds.), Marcel Dekker, Inc., New York, 1986.

Separation of Bioproducts

Cell culture bioreactors produce a dilute mixture of cells in an aqueous medium. Recovery of the product proteins from these cells may require disruption of the cells. This creates a host of problems. Cell walls and organelles must be removed. Proteins must be concentrated from a highly dilute solution that is mostly composed of water and other small molecules. The desired proteins must be separated from other macromolecules with similar physical properties. For biologically active proteins, separations must not only be specific for the target proteins, but also gentle enough to prevent denaturation and loss of biological activity and suitable for large-scale operation. Solving these

problems requires generic research on highly selective separations, as well as on the problems of concentrating materials from very dilute solutions (Figure 3.5). These and other generic research opportunities in separations have been described in detail in a recent report from the National Research Council.[1] Pursuing these opportunities will result in a better understanding of separation processes now used for the large-scale purification of proteins (e.g., precipitation and process chromatography). It may also result in novel separations involving aspects of techniques such as

- chromatography,
- membrane separation,
- fractionation in electric and gravitational fields,
- immunoadsorption,
- extraction with supercritical fluids,
- two-phase aqueous solution extraction, and
- separation by use of microemulsions.

The development of such new separations is crucial to the development of industrial biotechnology.

Another approach to separation problems lies in the development of modified organisms that produce the target proteins in high yield and concentration, thus reducing the time and cost of separating the proteins from large amounts of water. This is an area where early involvement of chemical engineers in designing genetically engineered organisms would be valuable. With their insights into the requirements of downstream processing of biologically synthesized substances, chemical engineers could be valuable members of an interdisciplinary team of molecular biologists and biochemists seeking to tailor the genetic code of cells.

Engineering Analysis of Complex Biological Systems

The development of new therapeutic procedures will be aided by a better understanding of physiological and pathological processes in the body. One area to which chemical engineers can contribute is the application of engineering analysis to systems found in the body. The

FIGURE 3.5 The desired product, usually a protein, produced by a genetically engineered microorganism must be separated and purified before use. The centrifuges shown here separate the components of the microorganisms, and then further separation is carried out to isolate the one protein that is desired from the thousands of other proteins produced by the microorganism. The isolated protein must be rigorously purified to eliminate contaminants from the final product. In many cases, separation and purification is the most expensive part of the production process. Courtesy, Genentech.

study of the transport of substances across membranes is an example. There is considerable knowledge of the transport of small molecules across living membranes; this should be extended to studies of larger molecules. A more complete understanding of the transport of biologically active agents would be particularly important in diagnosis and therapy.

Biochemical processes in humans can now be measured by such techniques as positron emission tomography, magnetic resonance imaging, and x-ray computer-assisted tomography, and the measurements can be enhanced by digital subtraction methods. Chemical engineers can help elucidate the data obtained by such techniques by developing quantitative models that incorporate thermodynamics, transport phenomena, fluid mechanics, and principles of chemical reaction engineering. These advances will lead to improved therapeutic procedures.

The normal growth of tissues and organs is under a remarkable degree of natural control. When this is compromised by genetic or mu-tagenic alterations, pathological processes such as birth defects or cancer can result. We need a better basic understanding of this control process. Theoretical and systematic advances by chemical engineers in process control may be applicable to the study of this problem.

While the mechanical performance of artificial materials in the human body can be predicted with some reliability, forecasting their biological performance is difficult. The problem of interactions at surfaces has already been mentioned. Research frontiers also include developing ways to simulate in vivo processes in vitro and extending the power and applicability of such simulations to allow for better prediction of the performance of biomedical materials and devices in the patient. Fundamental information on the correlation between the in vivo and in vitro responses is limited. Chemical engineers might also make contributions to the problem of noninvasive monitoring of implanted materials.

IMPLICATIONS OF RESEARCH FRONTIERS

The most successful efforts on problems such as those listed above will come from a new breed of chemical engineer, fluent in the language and concepts of modern biology and medicine. Currently, few chemical engineers are sufficiently knowledgeable in the principles of modern molecular biology, microbiology, genetics, and biochemistry to permit their effective collaboration with life scientists. Conversely, few life scientists are sufficiently aware of the engineering principles and practical problems associated with the scale-up of biological processes, the large-scale processing of bioproducts, or the development of artificial biological devices. All the participating disciplines must recognize the importance of the innovative synthesis of new concepts that unite life science theory and fact with engineering principles, or that combine an engineering idea with a biological speculation. Such innovative synthesis is likely to come about only in an environment where research needs and unsolved problems can be identified that bridge disciplinary boundaries and compel representatives of all relevant disciplines to work together to find the best solutions. Prompt and effective exploitation of the ''new'' biology is dependent on the improvement of this disciplinary interface; and this is one of the most critical problems confronting bioengineering today.

The need to develop a new fusion with modern biology has important implications for chemical engineering education and research:

• The development of fruitful education and research programs in biochemical and biomedical engineering cannot take place in isolation from the life sciences; strong, complementary academic programs in the biological or medical sciences are essential. Institutions that do not have strong research activities in the life sciences should probably not be encouraged to develop programs in biochemical or biomedical engineering.

• Curricula at the undergraduate and graduate levels need to be modified so that students will gain sufficient knowledge of the biological sciences to apply engineering methods of analysis and design to solve problems that originate in the biological sciences. Chapter 10 discusses general principles for modifying the undergraduate curriculum to respond to emerging applications for chemical engineering. At the graduate level, in-depth courses in molecular biology, biochemistry, and cellular and mammalian

Understanding Kidney Function Through Chemical Engineering Research

The kidneys play a critical role in regulating the composition and volume of blood and other body fluids. In doing so, they help ensure a healthy internal environment in the body. The first step in urine formation is the filtration of blood across the walls of certain microscopic blood vessels in the kidney. This filtration process allows some of the water and low-molecular-weight substances dissolved in blood plasma to enter the renal tubules, while retaining blood cells and essentially all plasma proteins within the bloodstream. The blood vessels involved, called glomerular capillaries, are arranged in spherical tufts a fraction of a millimeter in diameter, called glomeruli. Injury to many of the million or so glomeruli present in each kidney is a frequent feature of renal disease.

During the early 1970s, technical innovations by renal physiologists permitted the first direct measurements of the pressures responsible for glomerular filtration in mammals. This created the opportunity to use engineering principles in analyzing kidney function. The resulting collaboration between physiologists and chemical engineers led to several major new insights.

Hydraulic permeability—the rate of filtration per unit of applied pressure—was one of the physical properties that were studied. It was discovered that the hydraulic permeability of the glomerular capillary wall was twice that of capillaries in other organs. This surprising finding implied that the limiting factor in kidney function was not hydraulic permeability, as had been assumed, but rather the rate of plasma flow to the kidney. It was further discovered that the design of the glomerulus ingeniously avoided one of the classic problems of man-made filtration processes: concentration

physiology should be part of the course requirements for chemical engineers specializing in bioengineering. Such courses should be structured specifically for engineers, include meaningful laboratory experience, and provide the prerequisite background for the engineering student to take advanced biology and medical science courses, if desired.

• Ph.D. students must be prepared for the interdisciplinary environment in which they will likely spend their careers as biochemical or biomedical engineers. The best way to do this is to expose them to interdisciplinary research as graduate students. To facilitate this, a broad and stable base of research support targeted at interdisciplinary research must be created. Particularly valuable would be support targeted to polarization, or the buildup at the surface of a filter of substances too large to pass through.

The glomeruli normally produce large volumes of filtrate without significant leakage of plasma proteins into the urine. The amount of fluid that is filtered by the glomeruli each day is some 50 times the volume of blood plasma. Because the body cannot rapidly replace plasma proteins, even small defects in the glomerular barrier are potentially disastrous. Prior to the involvement of chemical engineers in kidney studies, it was thought that protein leakage was prevented because the holes in the glomerular membrane were smaller than the major protein molecules, so that the capillary wall acted as a sieve. While sieving is one important part of glomerular function, further collaborative studies by physiologists and chemical engineers revealed another, unexpected phenomenon. The glomerulus was found to discriminate among circulating proteins on the basis of their electrical charge, in addition to their molecular size. Negatively charged molecules were repelled by negatively charged components in the capillary wall, and thus were retained in the bloodstream more effectively than uncharged molecules of the same size. The negatively charged molecules attached to the capillary walls of the glomeruli had been identified previously, but their functional significance was not appreciated.

These studies led to the realization that proteinuria—the abnormal appearance of protein in the urine—could result not only from the enlargement of submicroscopic holes in the glomerular capillary wall, but also from the loss or neutralization of its negatively charged components. This finding has provided a new direction for research on the molecular basis for the nephrotic syndrome, a group of kidney diseases all characterized by massive proteinuria.

medium-sized research collaborations bringing together two or three co-principal investigators whose backgrounds and expertise cross the boundary between chemical engineering and the life sciences, including medicine. (See "Cross-disciplinary partnership awards" in Chapter 10.) While large centers certainly can provide an interdisciplinary research environment, a greater number of medium-sized collaborations might foster a faster growth of U.S. capabilities in critical bioengineering areas.

• A faculty expert in both the engineering and the biological aspects of the research frontiers described in this chapter is needed to mount a significant educational program in biochemical and biomedical engineering. The hiring of faculty into chemical engineering departments whose training is initially in the medical and life sciences is one step that might be encouraged. The presence of a strong research biology department or a nearby teaching hospital/medical school is probably needed to furnish an environment that will attract and retain the best such faculty. There are many practical obstacles to be overcome in making such appointments successful. A program to encourage "pioneers" who wish to cross the disciplinary divide into chemical engineering departments is outlined in Chapter 10. Some bioengineering departments have already made joint appointments with biological and medical faculties. Where the organizational problems inherent in such arrangements can be avoided or resolved, such appointments should be encouraged.

A number of other factors will be important in sustaining a vital research effort in biochemical and biomedical engineering. These include:

• *Instrumentation and facilities.* Suitable instrumentation and facilities for education and research in bioengineering can be very expensive. For example, equipping a state-of-the-art tissue-culture facility for engineering studies costs in the range of $500,000. Other costly equipment required includes ultracentrifuges, electron microscopes, mass spectrometers, NMR spectrometers, scintillation counters, and specialized instruments to study surfaces (see Chapter 9). Some of this equipment must be specially modified and dedicated to a particular group's use. Other instruments can be shared among a coterie of chemical engineering, biological, and medical researchers. Chemical engineers should make use of existing facilities in life sciences and medical departments wherever possible, particularly in the case of animal facilities.

● *National research centers.* Special, highly sophisticated ensembles of analytical and computational equipment and expertise might be brought together in national research facilities available to academic, government, and industrial groups for limited time periods. Some potential areas of specialization for such centers include modeling and control of bioreactors, measuring and modeling pharmacokinetic data, measuring actual and simulated bioflows in living systems and reactors, and studying kinetics of biological reactions and related processes.

● *Effective coupling to industry.* Effective links between universities and industry are essential to successful research and education in biochemical and biomedical engineering. In this rapidly growing technological area, a particular need is effective contact and interchange between chemical engineering departments and smaller venture-capital firms specializing in biotechnology or biomedical products. Liaison programs and other mechanisms that promote interactions between active researchers, and opportunities for students to spend time in industrial laboratories, should be encouraged.

● *Better communication among professional societies.* The field of biochemical and biomedical engineering is in danger of fragmentation among a plethora of professional societies, some of which are quite narrow in focus. Literally dozens of such organizations are currently on the scene. The AIChE could play a valuable role in ameliorating this situation by promoting better communication and cooperation among societies and researchers in other disciplines.

The biochemical and biomedical engineers of the future will be in great demand by industry, academia, and federal and state government agencies. Already, there is a strong demand by universities for faculty in biochemical and biomedical engineering. While recent demand from industry has not been as intense, it is projected to increase strongly as products are better defined and move closer to commercial production.[2] Federal and state agencies that will be responsible for regulating the introduction of new bioproducts into society are woefully understaffed in biologically conversant engineers. These agencies (e.g., EPA, USDA, and FDA) should also support chemical engineering research to obtain the data, models, and insight necessary for effective risk assessment and management.

It is characteristic of U.S. labor markets for scientific and engineering personnel to experience severe shortages and overcompensating excesses. Now is the time for the federal government and universities to build a research and education base in academia that can respond flexibly and efficiently to the personnel demands that will inevitably come. Now is the time to prepare a cadre of chemical engineers who will interact as easily and successfully with life scientists as chemical engineers currently do with chemists and physicists.

NOTES

1. National Research Council, Committee on Separation Science and Technology. *Separation and Purification: Critical Needs and Opportunities.* Washington, D.C.: National Academy Press, 1987.
2. U.S. Congress, Office of Technology Assessment. *Commercializing Biotechnology—An International Analysis.* Washington, D.C.: U.S. Government Printing Office, 1983.

SUGGESTED READING

J. Feder and W. R. Tolbert. "The Large-Scale Cultivation of Mammalian Cells." *Sci. Am.*, 248 (1), January 1983, 36.

E. L. Gaden, Jr. "Production Methods in Industrial Microbiology." *Sci. Am.*, 245 (3), September 1981, 180.

A. E. Humphrey. "Commercializing Biotechnology: Challenge to the Chemical Engineer." *Chem. Eng. Prog.*, 80 (12), December 1984, 7.

A. S. Michaels. "Adapting Modern Biology to Industrial Practice." *Chem. Eng. Prog.*, 80 (6), June 1984, 19.

A. S. Michaels. "The Impact of Genetic Engineering." *Chem. Eng. Prog.*, 80 (4), April 1984, 9.

National Academy of Sciences-National Academy of Engineering-Institute of Medicine, Committee on Science, Engineering, and Public Policy. "Report of the Research Briefing Panel on Chemical and Process Engineering for Biotechnology," in *Research Briefings 1984.* Washington, D.C.: National Academy Press, 1984.

National Research Council, Engineering Research Board. "Bioengineering Systems Research in the United States: An Overview," in *Directions in Engineering Research*. Washington, D.C.: National Academy Press, 1987.

National Research Council, National Materials Advisory Board. *Bioprocessing for the Energy-Efficient Production of Chemicals*. Washington, D.C.: National Academy Press, 1986.

R. A. Weinberg. "The Molecules of Life." *Sci. Am.*, 253 (4), October 1985, 48.

Electronic, Photonic, and Recording Materials and Devices

The information technologies on which modern society depends would not be possible without integrated circuits, optical fibers, magnetic media, devices for electrical interconnection, and photovoltaics. In the future, these technologies may undergo another revolution as superconductors find use in devices for information storage and handling. Chemical processes are the means by which the physical properties and structural features of these materials and devices are established and tailored. Chemical engineers are beginning to play an important role in process design, optimization, and control in the electronics industry. Their contribution to improving manufacturing technology is particularly welcome as the United States faces fierce international competition from Japan and Korea. These two countries have already gained the edge in some manufacturing areas and are making inroads into U.S. markets for electronics and recording devices. This chapter describes intellectual frontiers for chemical engineers that, if successfully pursued, can improve the U.S. competitive position. These frontiers include process integration, reactor design and engineering, ultrapurification, materials synthesis and processing, thin film deposition, modeling, chemical dynamics, and process design and control for safety and environmental protection. Implications of these challenges for the profession are discussed.

ALMOST EVERY ASPECT of our lives—at work, at home, and in recreation—has been affected by the information revolution. Today, information is collected, processed, displayed, stored, retrieved, and transmitted by an array of powerful technologies that rely on electronic microcircuits, light wave communication systems, magnetic and optical data storage and recording, and electrical interconnections. Materials and devices for these technologies, along with photovoltaic materials and devices, are manufactured by sophisticated chemical processes. The United States is now engaged in fierce international competition to achieve and maintain leadership in the design and manufacture of these materials and devices. The economic stakes are large (see Table 4.1); national productivity and security interests dictate that we make the strongest possible effort to stay ahead in processing science and technology for this area.

In the manufacturing of components for information and photovoltaic systems, there has been a long-term trend away from mechanical production and toward production by chemical processes. Chemists and chemical engineers have become increasingly involved in several areas of research and process development. Worldwide, however, many high-technology industries, such as microelectronics, still have surprisingly little strength in chemical processing and engineering. The United States has a particular advantage over its international competitors in that its chemical engineering research community leads the world in size and sophistication. The United States is in a position to exploit its strong competence in chemical processing to regain leadership in areas where the initiative in manufacturing technology has passed to Japan and to maintain or increase leadership in areas of U.S. technological strength.

Table 4.2 illustrates some of the ways in which chemical engineering can contribute to research on information and photovoltaic materials and devices. To fully achieve its potential contribution, though, the field of chemical engineering must strongly interact with other disciplines in these industries. Chemical engineers must be able to communicate across disciplinary lines, as the technologies discussed in this chapter involve solid-state physics and chemistry, electrical engineering, and materials science.

Electronic, photonic, and recording materials and devices may seem to be an exceedingly diverse class of materials, but they have many characteristics in common: their products are high in value; they require relatively small amounts of energy or materials to manufacture; they have short commercial life cycles; and their markets are fiercely competitive—consequently, these products experience rapid price erosion. The manufacturing methods used to produce integrated circuits, interconnections, optical fibers, recording media, and photovoltaics also share characteristics. All involve a sequence of individual, complex steps, most of which entail the chemical modification or synthesis of materials. The individual steps are designed as discrete unit or batch operations and, to date, there has been little effort to integrate the overall manufacturing process. Chemical engineers can play a significant role in improving manufacturing processes and techniques, and investments in chemical processing science and engineering research represent a potentially high-leverage approach to enhancing our competitive position.

CURRENT CHEMICAL MANUFACTURING PROCESSES

Before the invention of the transistor in 1948, the electronics industry was based on vacuum

TABLE 4.1 Worldwide Market for Materials and Devices for Information Storage and Handling (billions of 1986 dollars)

Technology	Year		
	1985	1990[a]	1995[a]
Electronic semiconductors	25	60	160
Light wave fiber and devices	1	3	5.5
Recording materials	7	20	55
Interconnections	10	21	58
Photovoltaics	0.3	0.8	3
Total electronics	397	550	n.a.

[a] Market projection.

SOURCE: AT&T Bell Laboratories. Compiled from various published sources.

TABLE 4.2 Chemical Engineering Aspects of Electronic, Photonic, and Recording Materials and Devices

Chemical Engineering Contributions	Technologies				
	Microcircuits	Photovoltaic Devices	Optical and Magnetic Storage and Recording	Light Wave Media and Devices	Interconnection
Large-scale synthesis of materials	Ultrapure single-crystal silicon III-V compounds II-VI compounds Electrically active polymers Other chemicals used in processing of electronic materials	Ultrapure amorphous silicon III-V compounds II-VI compounds	Optical recording materials E-beam recording materials Magnetic particles Film and disk substrates	Ultrapure glass Fiber coatings	Bonding materials Polyimides Reinforced composites Ceramic substrates
Engineering of reaction and deposition processes	Chemical vapor deposition Physical vapor deposition Plasma vapor deposition Wet chemical etching Plasma etching	Chemical vapor deposition Physical vapor deposition Plasma vapor deposition Wet chemical etching Plasma etching	Sputtering Plasma-enhanced chemical vapor deposition Plating Solution coating	Modified chemical vapor deposition Sol-gel processing	Plating
Other challenges in engineering chemical processes	Effluent treatment	Effluent treatment		Fiber drawing and coating processes	Ceramic processing
Packaging and/or assembly	Process integration and automation New materials for packaging Modeling of heat transfer in design of packaging	Process integration and automation	Process integration	Cabling Laser packaging	Wafer-scale automation

tube technology, and most electronic gear was assembled on a metal chassis with mechanical attachment, soldering, and hand wiring. All the components of pretransistor electronic products— vacuum tubes, capacitors, inductors, and resistors—were manufactured by mechanical processes.

A rapid evolution occurred in the electronics industry after the invention of the transistor and the monolithic integrated circuit:

• Today's electronic equipment is filled with integrated circuits, interconnection boards, and other devices that are all manufactured by chemical processes.

• The medium used for the transmission of information and data over distances has evolved from copper wire to optical fiber. It is quite likely that no wire-based information transmission systems will be installed in the future. The manufacture of optical fibers, like that of microcircuits, is almost entirely a chemical process.

• Early data storage memory was based on ferrite core coils containing a reed switch that mechanically held bits of information in either an on or off state. Today, most data is permanently stored through the use of magnetic materials and devices, and the next generation of data storage devices, based on optoelectronic materials and devices, is beginning to enter the marketplace. Ferrite cores were manufactured by winding coils and mechanically mounting the individual memory cells in large arrays. Magnetic and optical storage media are manufactured almost entirely by chemical processes.

The importance and sophistication of current chemical manufacturing processes for electronic, photonic, and recording materials and devices are not widely appreciated. A more detailed description serves to highlight their central role in these technologies.

Microcircuits

A semiconductor microcircuit is a series of electrically interconnected films that are laid down by chemical reactions. The successful growth and manipulation of these films depend heavily on proper design of the chemical reac-

tors in which they are laid down, the choice of chemical reagents, separation and purification steps, and the design and operation of sophisticated control systems. Microelectronics based on microcircuits are commonly used in such consumer items as calculators, digital watches, personal computers, and microwave ovens and in information processing units that are used in communication, defense, space exploration, medicine, and education.

Microcircuitry has been made possible by our ability to use chemical reactions and processes to fabricate millions of electronic components or elements simultaneously on a single substrate, usually silicon. For example, a 1-million-bit dynamic random access memory device (Figure 4.1) contains 1.4 million transistors and 1 million capacitors, with some chemically etched features on the chip being as small as 1.1 μm. This stunning achievement is just one step in a long-term trend toward the design and production of integrated circuits of increasing complexity and capability. There is still considerable room for further increases in component density in silicon-based microelectronics (Figure 4.2), not to mention possible advances in component density that would result from alternative meth-

FIGURE 4.1 Chemical reactions are used to achieve the fine structures seen in modern integrated circuits. This electron micrograph shows a transistor in a "cell" of a 1-megabit dynamic random access memory chip. The distance between features is about 1 μm. Courtesy, AT&T Bell Laboratories.

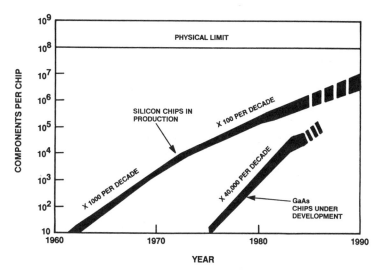

FIGURE 4.2 The chemical processes used for the manufacture of microcircuits have become progressively more sophisticated. This development is responsible for the large increases in the number of components that can be placed on a single chip. Trends in increasing component density are shown from 1960 for silicon chips (top line) and from 1975 for developmental chips based on gallium arsenide (bottom line). The rates of growth shown have been remarkable. From 1962 to 1972, silicon component density increased a thousandfold and from 1972 to 1982, a hundredfold. From 1975 to 1985, component density in developmental gallium arsenide devices grew by a factor of 40,000. Courtesy, AT&T Bell Labortories.

ods of storing and transferring information (e.g., three-dimensional circuits and Josephson (quantum) or optical devices).

Chemical reactions and processes in the manufacture of microcircuits (Figure 4.3) begin with the basic material for integrated circuits, high-purity (less than 150 parts per trillion of impurities) polycrystalline silicon. This ultrapure silicon is produced from metallurgical grade (98 percent pure) silicon by the following steps (Figure 4.4):

• reaction at high temperature with hydrogen chloride to form a complex mixture containing trichlorosilane;

• separation and purification of of trichlorosilane by absorption and distillation; and

• reduction of ultrapure trichlorosilane to polycrystalline silicon by reaction with hydrogen at 1,100–1,200°C.

To prepare single-crystal silicon ingots suitable for use as materials in semiconductors, polycrystalline silicon is melted in a crucible at 1,400–1,500°C under an argon atmosphere. Tiny quantities of dopants—compounds of phosphorus, arsenic, or boron—are then added to the melt to achieve the desired electrical properties of the finished single-crystal wafers. A tiny seed crystal of silicon with the proper crystalline orientation is inserted into the melt and slowly rotated and withdrawn at a precisely controlled rate, forming a large cylindrical single crystal 6 inches (14 cm) in diameter and about as tall as an adult human being (1.8 m) with the desired crystalline orientation and composition. Crystal growth kinetics, heat and mass transfer relationships, and chemical reactions all play important roles in this process of controlled growth. The resulting single-crystal ingots are sawed into wafers that are polished to a flatness in the range of from 1 to 10 μm.

The next steps in device fabrication are the sequential deposition and patterning of thin dielectric and conducting films (Figure 4.5). The polished silicon wafer is first oxidized in a furnace at 1,000–1,200°C. The resultant silicon dioxide film is a few hundred nanometers thick and extremely uniform. The wafer is then coated with a photosensitive polymeric material, termed a resist, and is exposed to light through the appropriate photomask. The purpose of the photolithographic process is to transfer the mask pattern to the thin film on the wafer surface. The exposed organic film is developed with a solvent that removes unwanted portions, and the resulting pattern serves as a mask for chemically etching the pattern into the silicon dioxide film. The resist is then removed with an oxidizing agent such as a sulfuric acid-hydrogen peroxide mixture, and the wafer is chemically cleaned and ready for other steps in the fabrication process.

The patterned wafer might next be placed in a diffusion furnace, where a first doping step is performed to deposit phosphorus or boron into

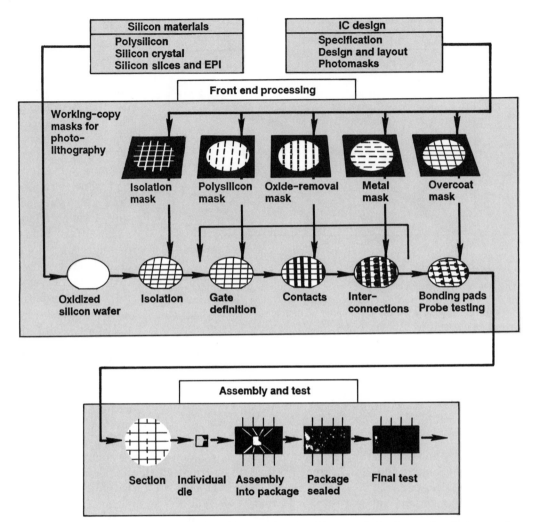

FIGURE 4.3 The manufacture of integrated circuits requires both expertise in electronic design and chemical processing. Chemical process steps are important to the preparation of silicon materials, to the steps from oxidation of silicon wafers through establishment of bonding pads, and to the final assembly of chips in individual packages. Excerpted by special permission from *Chemical Engineering*, June 10, 1985. Copyright 1985 by McGraw-Hill, Inc., New York, NY 10020.

the holes in the oxide. A new oxide film can then be grown and the photoresist process repeated. As many as 12 layers of conductor, semiconductor, and dielectric materials are deposited, etched, and/or doped to build the three-dimensional structure of the microcircuit.

Light Wave Media and Devices

Photonics involves the transmission of optical signals through a guiding medium—generally a

glass fiber—for purposes that include telecommunications, data and image transmission, energy transmission, sensing, display, and signal processing. Optical fiber technology is less than 14 years old and only became a commercial reality in the early 1980s. It is now a $1 billion per year industry. The data-transmitting capacity of optical fiber systems has doubled every year since 1976 (Figure 4.6). In fact, optical fiber systems planned on the basis of the prevailing technology at that time are often obsolete

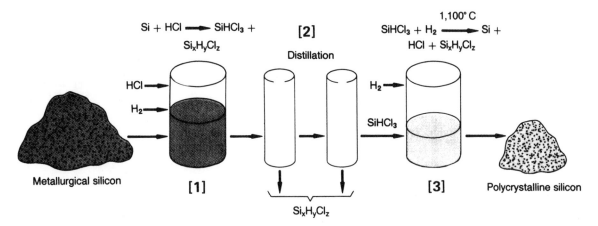

FIGURE 4.4 The production of polycrystalline silicon for the electronics industry involves several chemical steps aimed at the reduction of impurities. These include (1) reaction of metallurigcal grade silicon to produce a mixture of chlorosilanes, (2) distillation of trichorosilane, and (3) reduction of trichloro-silane to polycrystalline silicon. Excerpted by special permission from *Chemical Engineering*, June 10, 1985. Copyright 1985 by McGraw-Hill, Inc., New York, NY 10020.

by the time they are implemented. Typically, a given light wave technology is supplanted by an improved technology after one year.

Other applications for light guides, such as optical fiber sensors and transducers, are receiving a great deal of attention. Image transmission (e.g., endoscopes), energy transmission (e.g., light pipes), and display (e.g., decorative signs) are growing commercial areas.

Light wave media and devices include the guiding medium (optical fibers), sending and receiving devices, and associated electronics and circuitry. The transmission of light signals through optical fibers must occur at wavelengths where the absorption of light by the fiber is at a minimum. Typically, for SiO_2/GeO_2 glass, the best transmission windows are at 1.3 or 1.5 μm (Figure 4.7).

Optical signal processing for integrated optics and optical computing is in a rudimentary state but is certain to be an important area for future technological development. The processing involved in making optoelectronic devices is very similar to that used in microcircuit manufacture, but with considerable utilization of Group III-V compound semiconductors, lithium niobate, and a variety of polymeric materials. Developmental manufacturing processes for optoelec-

tronics emphasize reactive ion etching, epitaxy (e.g., metalorganic chemical vapor deposition (MOCVD), vapor-phase epitaxy, and molecular-beam epitaxy (MBE)), and photochemical and beam processing techniques for writing circuit configurations. All these processes are based on chemical reactions that require precise process control to produce useful devices.

Optical fibers are made by chemical processes. The critical feature of an optical fiber that allows it to propagate light down its length is a core of high refractive index surrounded by a cladding of lower index. The higher index core is produced by doping silica with oxides of phosphorus, germanium, and/or aluminum. The cladding is either pure silica or silica doped with fluorides or boron oxide.

There are four principal processes that may be used to manufacture the glass body that is drawn into today's optical fiber. "Outside" processes—outside vapor-phase oxidation and vertical axial deposition—produce layered deposits of doped silica by varying the concentration of $SiCl_4$ and dopants passing through a torch. The resulting "soot" of doped silica is deposited and partially sintered to form a porous silica boule. Next, the boule is sintered to a pore-free glass rod of exquisite purity and trans-

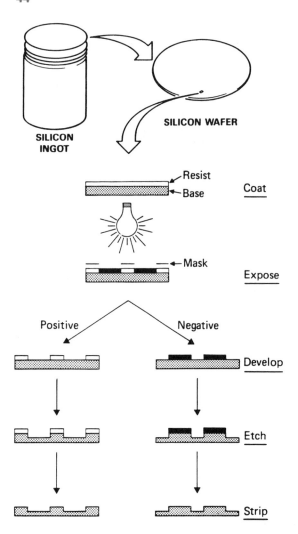

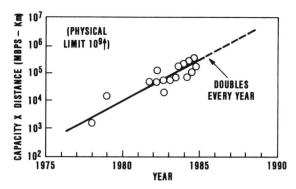

FIGURE 4.6 Since 1975, both the capacity of optical fiber and the distance a signal can be carried on optical fiber have steadily increased. Courtesy, AT&T Bell Laboratories.

parency. "Inside" processes—such as modified chemical vapor deposition (MCVD) and plasma chemical vapor deposition (PCVD)—deposit doped silica on the interior surface of a fused silica tube. In MCVD, the oxidation of the halide reactants is initiated by a flame that heats the outside of the tube (Figure 4.8). In PCVD, the reaction is initiated by a microwave plasma. More than a hundred different layers with different refractive indexes (a function of glass composition) may be deposited by either process before the tube is collapsed to form a glass rod.

In current manufacturing plants for glass fiber, the glass rods formed by all the above processes are carried to another facility where they are

FIGURE 4.5 Chemical steps in photolithography. A simplified series of steps in photolithography is shown. A silicon wafer, taken from a single-crystal silicon ingot, is coated with a polymer resist that is sensitive to light. A mask is placed over the wafer and the resist is thus selectively exposed to light. Depending on the type of polymer coating used, two things can happen. If the polymer is a positive resist, exposure to light makes the polymer easier to dissolve in a solution during the development step. After the development step, a protective film is left on the wafer that is the image of the mask used. If the polymer is a negative resist, exposure to light makes the polymer more difficult to dissolve during the development step. Afterwards, a protective film is left on the wafer that is the opposite of the image on the mask used. A corrosive gas or liquid is then used to etch away those parts of the wafer unprotected by the resist film. The resist film is removed after etching in preparation for other process steps. Excerpted by special permission from *Chemical Engineering*, June 10, 1985. Copyright 1985 by McGraw-Hill, Inc., New York, NY 10020.

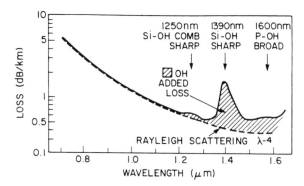

FIGURE 4.7 Transmission losses in glass fibers carrying optical signals are due to the interaction of light with chemical bonds. From 1.2 μm to 1.6 μm, losses due to Rayleigh scattering in the fiber are minimized, but transmission losses from Si-OH and P-OH bonds become large. The lowest transmission losses occur at wavelengths of 1.3 and 1.5 μm. Courtesy, AT&T Bell Laboratories.

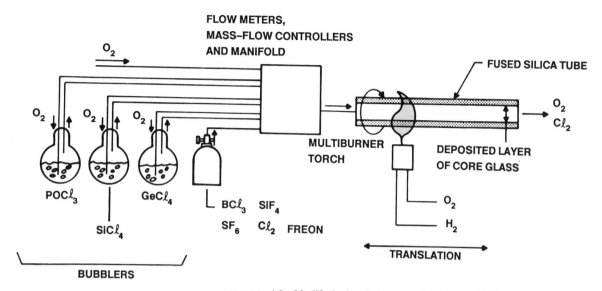

FIGURE 4.8 Modified chemical. vapor deposition (MCVD) is one of the principal processes used to manufacture optical fiber. In MCVD, a mixture of gases (O_2, $POCl_3$, $SiCl_4$, $GeCl_4$, BCl_3, SiF_4, SF_6, Cl_2, and freon) pass down the interior of a hollow silica tube that is being externally heated by a moving flame. The gases react to form a fine layer of silica glass doped with constituents of the gaseous mixture. Many layers can be deposited before the silica tube is collapsed and drawn into optical fiber. Courtesy, AT&T Bell Laboratories.

drawn into a thin fiber and immediately coated with a polymer. The polymer coating is important; it protects the fiber surface from microscopic scratches, which can seriously degrade the glass fiber's strength.

Current manufacturing technologies for optical fiber are expensive compared with the low cost of commodity glass. U.S. economic competitiveness in optical technologies would be greatly enhanced if low-cost means were found for producing wave guide-quality silica glass. The manufacture of glass lends itself to a fully integrated and automated (i.e., continuous) process. One can envision a fiber manufacturing plant that moves from purification of chemical reagents to a series of chemical reactions, glass-forming operations, and, finally, fiber-drawing steps. Intermediate products would never be removed from the production line. Sol-gel and related processes (see Chapter 5) are attractive candidates for such a manufacturing technology, which would start with inexpensive ingredients and proceed from a sol to a gel, to a porous silica body, to a dried and sintered glass rod, to drawn and coated fibers. Such a process

could reduce the cost of glass fiber by as much as a factor of 10, a step that would greatly increase the scope, availability, and competitiveness of light wave technologies.

At present the chemical steps involved in sol-gel processes are poorly understood. Methods are being sought to manipulate these processes to produce precisely layered structures in a reliable and reproducible way.

Recording Media

Recording media come in a number of formats (e.g., magnetic tape, magnetic disks, or optical disks) and are made by a variety of materials and processes (e.g., evaporated thin films or deposited magnetic particles in polymer matrixes). The next generation of recording media will be based on optical storage of data (Figure 4.9). Already, read-only optical disks (or CD-ROMs) are on the market for applications such as search and retrieval of information from large data bases. And optically based compact disks (CDs) are available in every record store. The possibility of creating optically based recording media for read-write storage of information has generated a tremendous amount of industrial research but, so far, no commercially viable products.

Since a practical read-write optical disk has not yet been invented, it is hard to describe the processing challenges involved in making it. Thus, the remainder of this section examines the most challenging (from a processing standpoint) of the remaining forms of recording media: magnetic disks and tape. Magnetic media are still an economically important part of the recording market and have a rich array of processing challenges with which chemical engineers have been involved. These challenges are relevant to the emerging technologies and materials in recording.

In the manufacture of magnetic recording media, the chemical and physical properties of the magnetic particles or thin films coated on a disk or tape are very important, for they determine the density at which information can be recorded. Paramount among these properties are the shape, size, and distribution of the magnetic particles. An extremely narrow size range of magnetic particles—themselves only a few tenths of a micrometer in diameter—must be achieved in a reliable and economic manner. Furthermore, the particles must be deposited in a highly oriented fashion and lie as closely together as possible, so that high recording densities can be achieved. To accomplish this, a variety of challenging problems must be solved in the chemistry and chemical engineering of barium ferrite and the oxides of chromium, cobalt, and iron (e.g., the synthesis and processing of micrometer-sized materials with specific geometric shapes).

The manufacture of magnetic tape illustrates an interesting sequence of chemical processing challenges (Figure 4.10). A carefully prepared dispersion of needle-like magnetic particles is coated onto a fast-moving (150–300 m/min) polyester film base 0.0066–0.08 mm thick. The ability to coat thin, smooth layers of uniform thickness is crucial. The particles, after being coated onto the film, are oriented in a desired direction either magnetically or mechanically during the coating process. After drying, the tape is calendered (squeezed between microsmooth steel and polymer rolls that rotate at different rates), providing a "microslip" action that polishes the tape surface. These manufacturing steps (i.e., materials synthesis, preparation and handling of uniform dispersions, coating, drying, and calendering) are chemical processes and/or unit operations that are familiar territory to chemical engineering analysis and design.

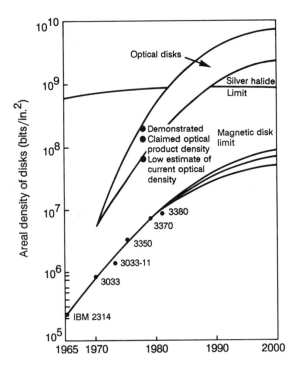

FIGURE 4.9 The density at which information can be written on optical disks (measured in bits/in²) was demonstrated in the early 1980s to be 10 times greater than the current highest performance magnetic disk. The gap between optical and magnetic storage capabilities is projected to increase over the next 20 years. Reprinted with permission from *Electronic Design*, August 18, 1983, 141.

Materials and Devices for Interconnection and Packaging

Interconnection and packaging allow electronic devices to be usefully incorporated into products. The manufacture of complex electronic systems requires that hundreds of thousands of electronic components be efficiently connected with one another in an extremely small space. In the past, this was accomplished by hand wiring discrete components on a chassis

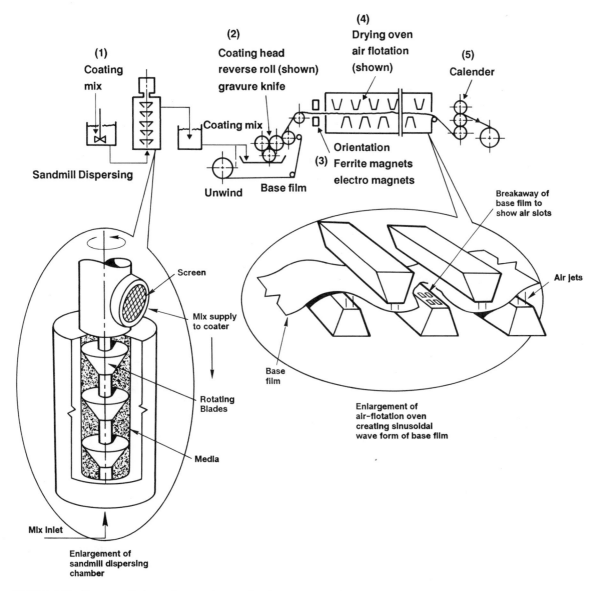

FIGURE 4.10 The manufacture of magnetic tape involves a series of steps including (1) forming a uniform dispersion of coating mix, (2) applying this coating to the film base, (3) orienting the magnetic particles, (4) drying the magnetic coat in an air-flotation oven, and (5) calendering and final wind-up on spools. Chemical processes are central to several of these steps. Excerpted by special permission from *Encyclopedia of Chemical Technology*, 3rd ed., Vol. 14, p. 745. Copyright 1978 by McGraw-Hill, Inc., New York, NY 10020.

assembly. Today, interconnection technology is based on high-density printed wiring boards, often with as many as 30 parallel layers of interconnection. The board insulation substrate may be either polymer or ceramic, with appropriate metal conductors.

The dielectric and the conductors are selected to maximize data transmission speed while minimizing signal loss. In addition, dissipating heat generated by the microcircuits is rapidly becoming an important consideration. If too much heat builds up in the microelectronic device, the device will start to fail. The next generation of interconnection substrates will likely require new materials and assembly techniques to cope with this challenge.

Plastic packaging of microelectronic devices is the most cost-effective means of providing electrical, mechanical, and environmental protection for an integrated circuit device. Approximately 80 percent of the billions of these devices used in the United States each year are packaged in plastic, and most are packaged in thermoset molding compounds by a conventional transfer molding process. This process must be carried out with exceptionally high yield, productivity, and reliability if the United States is to achieve a competitive advantage in the packaging of these devices.

All modern interconnection devices are manufactured by chemical processes such as etching, film deposition, and ceramic forming. Substrate formation is a vital part of the manufacturing process for fiberglass-impregnated printed wiring boards. It utilizes chemical processes such as metal deposition, lithographic patterning, etching, and chemical cleaning. Process design, to improve quality and decrease cost, continues to be a challenge, particularly in terms of greater uniformity in plating and etching and the environmental problems posed by the disposal of spent etching and plating baths.

Ceramic boards are currently widely used in high-performance electronic modules as interconnection substrates. They are processed from conventional ceramic precursors and refractory metal precursors and are subsequently fired to the final shape. This is largely an art; a much better fundamental understanding of the materials and chemical processes will be required if low-cost, high-yield production is to be realized (see

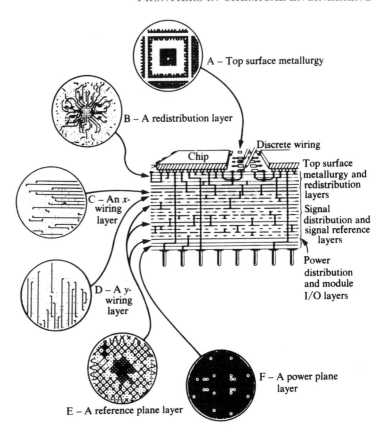

FIGURE 4.11 Cross-section of the IBM Multilayer Ceramic interconnection package. Various layers in this interconnection device are shown. Copyright 1982 by International Business Machines Corporation. Reprinted with permission.

Chapter 5). A good example of ceramic interconnection boards are the multilayer ceramic (MLC) structures used in large IBM computers (Figure 4.11). These boards measure up to 100 cm² in area and contain up to 33 layers. They can interconnect as many as 133 chips. Their fabrication involves hundreds of complex chemical processes that must be precisely controlled.

Better insulating substrates will be required for the ultrahigh-speed interconnection modules of the future. Although organic polymers appear to offer cost advantages for substrate application, the large-scale production of organic substrates with sufficiently low dielectric constants has not yet been realized. Achieving this will require the scale-up of new chemical reactions and the development of process methods for new polymer substrates.

Photovoltaics

A photovoltaic cell is a solid-state device that transforms solar energy into electricity. Significant research on photovoltaic cells began in the early 1970s and until now has generally focused on the invention and improvement of specific devices, including single-crystal silicon cells, amorphous silicon cells, heterojunction thin-film cells, and gallium arsenide cells.

Gallium arsenide cells have achieved high-energy efficiencies (in excess of 20 percent) and have been successfully used in systems where there is magnification of the sun's rays, that is, concentrator systems that increase the light flux per unit area. Substantial research is being devoted to improving the conversion efficiency and lowering the cost per kilowatt of such devices.

The materials and processes used in the manufacture of photoelectric energy conversion devices are almost identical to those used in manufacturing microelectronic devices and integrated circuits.

The photovoltaics industry could expand rapidly if the efficiency of polycrystalline modules could be increased to 15 percent, if these modules could be built with assurance of reliability over a 10- to 20-year period, and if they could be manufactured for $100 or less per square meter. Solar energy research has been largely directed toward only one of these issues: efficiency. All research aimed at reducing manufacturing costs has been done in industry and has been largely empirical. Almost no fundamental engineering research has been done on either the laboratory scale or the pilot plant scale for cost-effective processes for the production of photoconverters.

Superconductors

Superconductivity has been known since 1911, and superconducting systems based on various metal alloys (e.g., NbTi and Nb_3Sn) are currently used as magnets and in electronics. These materials exhibit superconductivity only at temperatures below 23 K and require cooling by liquid helium. The discovery of ceramics that exhibit superconductivity at temperatures up to 120 K, the so-called high-temperature superconductors, has sparked a tremendous amount of scientific activity and commercial interest around the world.

The key to the superconducting properties of these ceramics seems to be the presence of planes of copper and oxygen atoms bonded to one another. The significance of the other atoms in the lattice seems to be to provide a structural framework for the copper and oxygen atoms. Thus, in the superconducting compound YBa_2Cu_3O, the substitution of other rare earths for yttrium results in little change in the properties of the material.

Currently, superconducting materials are produced by standard techniques from the ceramics industry: mixing, grinding, and sintering. The basic structure of the 95 K superconductors is formed at temperatures above 800–900°C, and then annealed with oxygen at a temperature below 500°C. More needs to be known about the effect of various synthesis conditions on the microstructure found in these materials. Alternative routes to ceramics, such as sol-gel processes, may lead to significant improvements in the production of these materials. For microelectronic applications, various chemical vapor deposition routes to these materials need to be investigated. MOCVD might be a particularly promising route if volatile precursor compounds could be discovered and synthesized. Chemical engineers have the needed background for developing this technology (see Chapter 5) as well as finding the optimal procedure for drying, sintering, and calcining the final product.

INTERNATIONAL COMPETITION

In each of the technologies described in the preceding section, U.S. leadership in both fundamental research and manufacturing is severely challenged, and in some cases the United States is lagging behind its foreign competitors.

Microcircuits

A recent report of the National Research Council[1] has assessed the comparative position of the United States and Japan in advanced processing of electronic materials. The report, which focuses heavily on evaluating Japanese research on specific process steps in the manufacture of electronic materials, provides significant background for the following observations:

• The U.S. electronics industry appears to be ahead of, or on a par with, Japanese industry in most areas of current techniques for the deposition and processing of thin films—chemical vapor deposition (CVD), MOCVD, and MBE. There are differences in some areas, though, that may be crucial to future technologies. For example, the Japanese effort in low-pressure microwave plasma research is impressive and surpasses the U.S. effort in some respects. The Japanese are ahead of their U.S. counterparts in the design and manufacture of deposition equipment as well.

• Japanese industry has a very substantial commitment to advancing high-resolution lithography at the fastest possible pace. Two Japanese companies, Nikon and Canon, have made significant inroads at the cutting edge of optical lithography equipment. In the fields of x-ray and electron-beam lithography, it appears that U.S. equipment manufacturers have lost the initiative to Japan for the development of commercial equipment.

• Japanese researchers are ahead of their U.S. counterparts in the application of laser and electron beams and solid-phase epitaxy for the fabrication of silicon-on-insulator structures.

• The United States leads in basic research related to implantation processes and in the development of equipment for conventional applications of ion implantation. Japan appears to have the initiative in the development of equipment for ion microbeam technologies.

There is a penalty to be paid for falling behind foreign competitors in process equipment design and engineering. Early access to new prototypes of equipment allows a manufacturing firm to concurrently troubleshoot the equipment and integrate it into its existing process line. When the state-of-the-art processing equipment comes from overseas, companies in the country of

origin gain a competitive advantage stemming from this early access. A look at the installation record for JEOL focused ion beam instruments (which are the best in the world) illustrates this phenomenon (Table 4.3).

Much remains to be done, in both the United States and Japan, to solve the problems of process integration in microcircuit manufacture. Effort is being expended on equipment design for specific processing steps, but a parallel effort to integrate the processing of semiconductor materials and devices across the many individual steps has received less attention in both countries. Yet the latter effort may have significant payoffs in improved process reliability and efficiency—that is, in "manufacturability." The United States, with the strongest chemical engineering research community in the world, has the capability to take a significant lead in this area.

Light Wave Media and Devices

The Japanese are our prime competitors in the development of light wave technology. They are not dominant in the manufacture of optical fiber thanks in part to a strong overlay of patents on basic manufacturing processes by U.S. companies. In fact, a major Japanese company manufactures optical fiber in North Carolina for shipment to Japan. This is the only example to date of Japan importing a high-technology product from a U.S. subsidiary. Nonetheless, the Japanese are making strong efforts to surpass the United States and are reaching a par with us in many areas.

The United States still significantly leads Japan in producing special purpose and high-strength fibers, in preparing cables from groups of fibers, and in research on hermetic coatings for fibers.

Recording Media

Japan is the America's principal technological competitor in the manufacture of magnetic media, and Korean firms are beginning to make significant inroads at the low end of the magnetic tape market. U.S. companies producing magnetic tape use manufacturing processes that

TABLE 4.3   Installation Record for Focused Ion Beam Instruments Made by JEOL Semiconductor Equipment Division

Instrument Number	Customer	Country	Instrument Type	Year Installed
1	The Institute of Physical and Chemical Research	Japan	JIBL-34	1982
2	The Institute of Physical and Chemical Research	Japan	JIBL-100	1983
3	Optoelectronics Joint Laboratory	Japan	JIBL-100	1983
4	LSI R&D Lab, Mitsubishi Electric Corporation	Japan	JIBL-100	1983
5	Fujitsu Laboratories Ltd.–Atsugi	Japan	JIBL-100A	1984
6	Optoelectronics Joint Laboratory	Japan	JIBL-100A	1984
7	Institute of Laser Engineering, Osaka University	Japan	JIBL-30	1984
8	The Institute of Physical and Chemical Research	Japan	JIBL-200	1984
9	NTT Musashino Electrical Communication Laboratories	Japan	JIBL-100	1984
10	Institute of Industrial Science, Tokyo University	Japan	JIBL-100	1984
11	Fujitsu Laboratories Ltd.–Atsugi	Japan	JIBL-GP1	1984
12	Dainippon Screen	Japan	JIBL-100	1984
13	Fujitsu Laboratories Ltd.–Atsugi	Japan	JIBL-100	1984
14	NTT Atsugi Electrical Communication Laboratories	Japan	JIBL-100	1984
15	Tsukuba Research Center, Sanyo Electric Co., Ltd.	Japan	JIBL-100A	1984
16	NEC Corporation	Japan	JIBL-140	1984
17	LSI R&D Lab, Mitsubishi Electric Corporation	Japan	JIBL-140	1984
18	Optoelectronics Joint Laboratory	Japan	IPMA-10	1984
19	Institute of Industrial Science, Tokyo University	Japan	IPMA-10	1986
20	Matsushita Laboratory	Japan	JIBL-106	1986
21	Nihon Denso	Japan	JIBL-106	?
22	Sony	Japan	JIBL-GPI	?
23	Denka	Japan	JIBL-100	?
24	Max Planck Institute	Germany	JIBL-100A	?
25	IBM	U.S.A.	JIBL-106	1988

SOURCE: AT&T Bell Laboratories and JEOL.

achieve higher integration through combined unit operations, but Japanese companies have a higher degree of automation in these separate operations. U.S. companies lead the Japanese in the use of newer thermoplastics in calender-compliant roll materials. Japan used to surpass the United States in the product uniformity of magnetic tape for professional applications; U.S. firms have closed this gap in recent years and are now capturing worldwide market shares from the Japanese, even in Japan.

The most significant development in Japan is the entry of photographic film companies (Fuji and Konishuroku) into the manufacture of magnetic media. They are having a large impact because the heart of the manufacturing process is the deposition of thin layers, and chemical processing technology from the photographic film business can be used to improve the quality and yield of magnetic tape and disks.

The United States still lags behind Japan in the treatment and manufacture of magnetic particles (except possibly for 3M, which manufactures its particles internally). There are disturbing signs that the Japanese may be ahead of the United States in the next generation of film base, especially for vapor-deposition magnetic media. The situation is not entirely clear, because 3M and Kodak make their own proprietary film. Other U.S. magnetic media companies, though, may be buying their film technology from Japan in the future.

Optical recording media for read-write applications are still in the research stage. U.S. companies are roughly on par with European and Japanese companies in such research. Read-only applications (e.g., CD-ROM disks and compact audio disks) are largely dominated by manufacturing technology from overseas.

Interconnection and Packaging

The United States leads its competitors in the design of central processing unit packaging for large computers. Companies such as IBM, Cray, and Amdahl are on the cutting edge of interconnection design and manufacturing. Japanese companies are ahead in some interconnection technologies found in mid-sized and smaller computers (e.g., phenolic paper boards and epoxy-resin boards).

Photovoltaics

The U.S. photovoltaics industry serves more than 100 different countries. Major competition comes from Japan and, to a limited extent, from Europe. U.S. firms have a dominant position in the power module market (devices with photovoltaic areas greater than 0.5 m^2) while Japanese firms have dominated the consumer market for small-photovoltaic goods (e.g., calculators, watches, and radios).

Superconductors

A recent report on high-temperature superconductivity[2] characterizes international com-

petition as intense, but the U.S. competitive position in science as good. Japan, China, a number of European countries, and the USSR are putting in place significant scientific and technological efforts. In Japan, industrial consortia are being organized by the government to begin initial development activities. The report concludes, "Japan offers perhaps the strongest long-range competitive threat to the U.S. position."

General Observations

The industries that manufacture materials and components for information applications are characterized by products that are rapidly superseded in the market by improved ones. This rapid turnover stems from the intense competition among these industries and results in rapid price erosion for products, once introduced. These industries also require rapid technology transfer from the research laboratory onto the production line. Many of their *products* cannot be protected by patents, except for minor features. Therefore, the key to their competitive success is thoroughly characterized and integrated manufacturing processes supported by *process* innovations. In the past, much of the process technology on which these industries depend has been developed empirically. If the United States is to maintain a competitive position in these industries, it is essential that it develop the fundamental knowledge necessary to stimulate further improvement of, and innovation in, processes involving chemical reactions that must be precisely controlled in a manufacturing environment. In the next section, the principal technical challenges are set forth.

INTELLECTUAL FRONTIERS

A variety of important research issues require much more work if U.S. companies are to establish and maintain dominance in information storage and handling technologies. These issues are quite broad and cut across the spectrum of materials and devices.

Process Integration

Process integration is the key challenge in the design of efficient and cost-effective manufacturing processes for electronic, photonic, and recording materials and devices. Except for magnetic tape, these products are currently manufactured through a series of individual, isolated steps. If the United States is to retain a position of leadership, it is crucial that its overall manufacturing methodology be examined and that integrated manufacturing approaches be implemented. Historically, all industries have benefited both economically and in the quality and yield of products by the use of integrated manufacturing methods. As individual process steps become more complex and precise, the final results of manufacturing (e.g., yield, throughput, and reliability) often depend critically on the interactions among the various steps. Thus, it becomes increasingly important to automate and integrate individual process steps into an overall manufacturing process.

The concepts of chemical engineering are easily applied in meeting the challenge of process integration, particularly because many of the key process steps involve chemical reactions. For example, in the manufacture of microcircuits, chemical engineers can provide mathematical models and control algorithms for the transient and steady-state operation of individual chemical process steps (e.g., lithography, etching, film deposition, diffusion, and oxidation), as well as interactions between process steps and ultimately between processing and the characteristics of the final device. As another example, in microcircuit manufacture, chemical engineers can provide needed simulations of the dynamics of material movement through the plant and thus optimize the flow of devices (or wafers) through a fabrication line (Figure 4.12). The continuous production of photovoltaic devices will require similar studies with even more emphasis on automation.

Reactor Engineering and Design

Closely related to challenges in process integration are those in reactor engineering and design. Research in this area is important if we

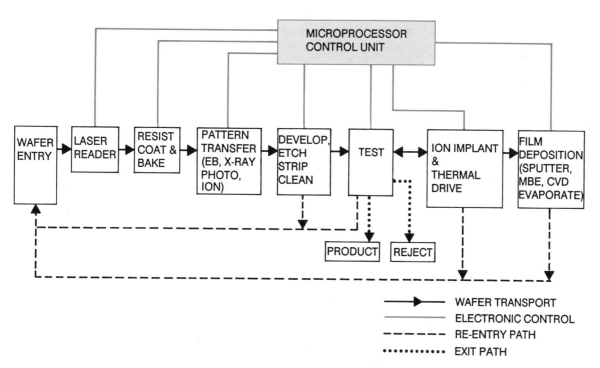

FIGURE 4.12 The integrated semiconductor processing line of the future will be a fully automated series of chemical processing steps. Chemical engineers will be needed to integrate individual process steps into a manufacturing line that can be operated free from human handling, and possible contamination, of the devices. Courtesy, AT&T Bell Laboratories.

are to automate manufacturing processes for higher yields and improved product quality. Processes such as CVD, epitaxy, plasma-enhanced CVD, plasma-enhanced etching, reactive sputtering, and oxidation all take place in chemical reactors. At present, processes and reactors are generally developed and refined by trial and error. A basic understanding of fundamental phenomena and reactor design would facilitate process design, control, and reliability. Because all these processes involve reaction kinetics, mass transfer, and fluid flow, chemical engineers bring a rich background to their study and improvement. For example, high-yield, continuous processes for film deposition and packaging are required if photovoltaic devices are to be manufactured at costs that are competitive with other energy technologies. New reactors and a better understanding of chemical dynamics in reactors are central to achieving this.

An important consideration in reactor design and engineering is the ultraclean storage and transfer of chemicals. This is not a trivial problem; generally, the containers and transfer media are the primary sources of contamination in manufacturing. Methods are needed for storing gases and liquids, for purifying them (see the next section), and for delivering them to the equipment where they will be used—all the while maintaining impurity levels below 1 part per billion. This requirement puts severe constraints on the types of materials that can be used in handling chemicals. For example, materials in reactor construction that might be chosen primarily on the basis of safety often cannot be used. Designs are needed that will meet the multiple objectives of high purity, safety, and low cost.

The ultimate limit to the size of microelectronic devices is of molecular dimensions. The ability to "tailor" films at the molecular level—to deposit a film and control its properties by altering or forming the structure, atomic layer by atomic layer—opens exciting possibilities for new types of devices and structures. The fabrication of these multilayer, multimaterial structures will require deposition methods such as MBE and MOCVD. Depositing uniform films by these methods over large dimensions will require reactors with a different design from those currently used, especially for epitaxial growth processes. The challenge is to be able to control the flow of reactants to build layered structures tens of atoms thick (e.g., superlattices). To achieve economic automated processes, the reactor design must allow for the acquisition of detailed real-time information on the surface processes taking place, fed back into an exquisite control system and reagent delivery system. This problem gives rise to an exciting series of basic research topics.

Ultrapurification

A third research challenge that is generic to electronic, photonic, and recording materials and devices stems from the need for starting materials that meet purity levels once thought to be unattainable.

This need is particularly acute for semiconductor materials and optical fibers. For semiconductor materials, the challenge is to find new, lower cost routes to ultrapure silicon and gallium arsenide and to purify other reagents used in the manufacturing process so that they do not introduce particulate contamination or other defects into the device being manufactured. For optical fibers, precursor materials of high purity are also needed. For example, the $SiCl_4$ currently used in optical fiber manufacture must have a total of less than 4 parts per million of hydrogen-containing compounds and less than 2 parts per billion of metal compounds (Figure 4.13). Either impurity will result in strong light absorption in the glass fiber. For magnetic media, the challenge is to separate and purify submicrometer-sized magnetic particles to very exacting size and shape tolerances.

A variety of separation research topics bear on these needs, such as generation of improved selectivity in separations by tailoring the chemical and steric interactions of separating agents, understanding and exploiting interfacial phenomena in separations, improving the rate and capacity of separations, and finding improved process configurations for separations. These are all research issues central to chemical engineering.

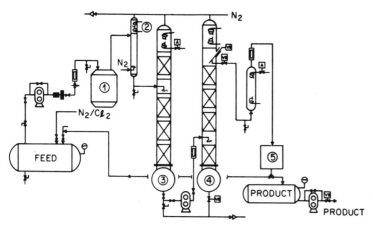

FIGURE 4.13 Schematic diagram for a purification plant for producing "optical fiber grade" $SiCl_4$. The feed material is passed through a reactor (1) where chlorination takes place. Excess HCl generated in the reactor is removed (2) and the product stream is passed through two distillation columns (3,4) where contaminants are removed. On-line IR spectroscopy (5) is used to monitor final product purity. Plants built using this design currently produce about 27 kg/h of ultrapure $SiCl_4$. Contaminants (e.g., compounds containing R-H, compounds containing C-H, and Fe) are reduced to below the limits of detection. Courtesy, AT&T Bell Laboratories.

Chemical Synthesis and Processing of Polymeric Materials

Although chemical engineering challenges related to polymeric materials are discussed in Chapter 5, the special challenges for polymers in materials and devices for information storage and handling deserve some mention here.

For the processing of microcircuits and interconnecting devices, improved radiation-sensitive polymers are needed for the formulation of better photoresists. Resists must be highly sensitive to the radiation used for exposure, but not to the microwave radiation used after development for other process steps such as plasma etching. Chemical engineering studies of polymer behavior during development steps are also needed. Details of the dissolution of the exposed (or unexposed) regions of the resist are at present poorly understood. There is a need for fundamental studies and modeling of the formation of a swollen gel layer at the solvent/polymer interface and the subsequent diffusion of polymer chains into solution.

Light wave technologies provide a number of special challenges for polymeric materials. Polymer fibers offer the best potential for optical

communications in local area network (LAN) applications, because their large core size makes it relatively cheap to attach connectors to them. There is a need for polymer fibers that have low losses and that can transmit the bandwidths needed for LAN applications; the acrylate and methacrylate polymers now under study have poor loss and bandwidth performance. Research on monomer purification, polymerization to precise molecular-size distributions, and well-controlled drawing processes is relevant here. There is also a need for precision plastic molding processes for mass production of optical fiber connectors and splice hardware. A tenfold reduction in the cost of fiber and related devices is necessary to make the utilization of optical fiber and related devices economical for local area networks and the telecommunications loop.

Another challenge for polymer research in light wave applications is in the use of active coatings on optical fibers as transducers for sensors. Such coatings may have magnetostrictive or piezoelectric properties. These coatings, or the fiber itself, may also incorporate dyes that would respond to chemicals, light, radiation, or other stimuli to produce transmission loss changes in the fiber. Such systems have enormous potential as sensors that would be ultrasensitive, capable of distributed sensing, able to operate in harsh environments, and unaffected by electromagnetic interference. Specialty fibers such as polarization-maintaining fibers, which have an asymmetric core and can double the bandwidth by transmitting two modes at once, may also play an important role in sensor technology.

Techniques for fabricating low-cost optical components such as graded index lenses, microlenses, couplers, splitters, and polarizers are needed to support optical fiber technology. Traditionally, amorphous inorganic materials have been used, but there are tremendous opportunities for innovation with polymers, which offer manufacturing versatility that is not available with glass. For example, photoselective polymerization techniques can be used to make branching wave guide circuits such as splitters and couplers. Photopolymerization and copolymerization of multiple monomer systems have been used to make radial, axial, and spherical graded-index lenses with a high degree of perfection (e.g., freedom from aberration). Large-scale, well-controlled chemical processes will be needed to fabricate these structures.

For recording applications, new approaches to high-quality polymeric film substrates are needed. Improved automation and control of thin-film coating are also important.

For interconnection and packaging technologies, an important goal is to achieve high-purity molding and dielectric materials. Epoxy-Novolac prepolymers with ionic impurity levels below 20 ppm offer one approach. There is a further need for low-viscosity molding compounds to minimize the development of flow stresses during processing. Continued development of thermally stable polymers with low dielectric constants (such as the polyimides) is also necessary. Advances in our fundamental understanding of polymer chemistry and rheology are crucial for all these areas (see Chapter 5).

Chemical Synthesis and Processing of Ceramic Materials

Challenges for chemical engineering related to ceramic materials are also discussed in Chapter 5, but the potential contribution of chemical engineers to this area cannot be emphasized too strongly. A tremendous opportunity exists for chemical engineers to apply their detailed knowledge of fundamental chemical processes in the development of new chemical routes to high-performance ceramics for electronic and photonic applications. The traditional approach to creating and processing ceramics has been through the grinding, mixing, and sintering of powders. Although still useful in many applications, this technology is being replaced by approaches that rely on chemical reactions to

create a uniform microstructure. Chemical routes to better ceramics have the advantage of being more amenable to continuous and automated processing. Among the typical examples of such approaches are sol-gel and related processes. (See Chapter 5 for a more detailed treatment of sol-gel processing.)

Deeper involvement of chemical engineers in manufacturing processes for ceramics may be particularly important to the eventual commercialization of metal oxide superconductors. The current generation of such superconductors consists of planar structures formed during a conventional ceramic synthesis. The ability to precisely control complex phase structure and phase boundaries seems critical. It is by no means clear that the formulations and structures that may produce optimal performance in superconducting ceramics (e.g., room-temperature superconductivity, capacity for high-current density) are accessible by these techniques. Rational synthesis of structured ceramics by chemical processing may be crucial to further improvements in superconducting properties and to efficient large-scale production.

Deposition of Thin Films

Precise and reproducible deposition of thin films is another area of great importance in the chemical processing of materials and devices for the information age.

In microelectronic devices, there is a steady trend toward decreasing pattern sizes, and by the end of this decade, the smallest pattern size on production circuits will be much less than 1 μm (Figure 4.14). Although the lithographic tools to print such patterns exist, the exposure step is only one of a number of processes that must be performed sequentially in a mass production environment without creating defects. Precise and uniform deposition of materials as very thin films onto substrates 14 cm or more in diameter must be performed in a reactor, usually at reduced pressure. Particulate defects larger than 0.1 μm must be virtually nonexistent. Low-temperature methods of film deposition will be needed so that defects are not generated in previous or neighboring films by unwanted diffusion of dopants.

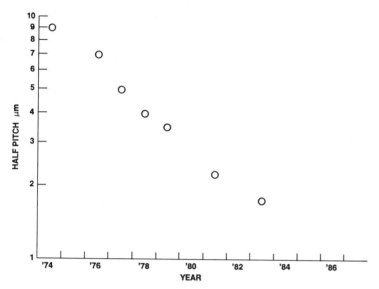

FIGURE 4.14 Feature size on microelectronic devices has steadily declined over the years as improved chemical etching processes have been developed. This graph shows feature size as a function of the year in which the device with the smallest feature size was first produced. Courtesy, AT&T Bell Laboratories.

For optical fibers, improved control over the structure of the thin films in the preform will lead to fibers with improved radial gradients of refractive index. A particular challenge is to achieve this sort of control in preforms created by sol-gel or related processes.

Another challenge in depositing thin films on optical fibers occurs in the final coating step. Improved coating materials that can be cured very rapidly, for example, by ultraviolet radiation, are needed for high-speed (>10 m/s) fiber-drawing processes. Both glassy and elastomeric polymers are needed for use over temperatures ranging from −60 to 84°C or higher. Hermetic coatings are required to avoid water-induced stress corrosion of silica glasses, which proceeds by slow crack growth. Materials under study include silicon carbide and titanium carbide applied by chemical vapor deposition, as well as metals such as aluminum. A tenfold increase in the rate at which such coatings can be applied to silica fiber during drawing is needed for commercial success. Coatings must be free of pinholes, have low residual stress, and adhere well. Hermetic coatings will also be needed to protect the moisture-sensitive halide

and chalcogenide glasses that may find use in optical fibers of the future because of their compatibility with transmission at longer wavelengths.

Considerable progress in the science and technology of depositing thin films is necessary if the U.S. recording media industry is to remain competitive with foreign manufacturers. New, fully automated coating processes that will generate high-quality, low-defect media are needed. Not only must considerable effort be mounted in designing hardware and production equipment, but complex mathematical models must be developed to study the kinetic and thermodynamic properties of film coating and the effect of non-Newtonian flow and polymer and fluid rheology. A better understanding of dispersion stability during drying, as well as of diffusion mechanisms that result in intermixing of sequential layers of macromolecules, is important.

Thin films are also critical to the performance of electrical interconnection devices (Figure 4.15). Better methods for depositing thin films conformably (for good sidewall coverage) and for achieving high-aspect-ratio trenches are needed for the interconnection of electronic devices for the high-frequency transmission of data. New processing strategies and device structures are required that use compatible layers of materials to minimize undesirable phenomena such as contact resistance; electromigration; leakage currents; delamination; and stress-related defects such as cracks, voids, and pinholes.

Modeling and the Study of Chemical Dynamics

A challenge related to the problems of reactor design and engineering is the modeling and study of the fundamental chemistry occurring in manufacturing processes for semiconductors, optical fibers, magnetic media, and interconnection.

For example, mathematical models originally developed for continuously stirred tank reactors and plug-flow reactors are applicable to the reactors used for thin-film processing and can be modified to elucidate ways to improve these reactors. For these models to reach their full descriptive potential, detailed studies of the fundamental chemical reactions occurring on surfaces and in the gas phase are required. For example, etching rates, etching selectivity, line profiles, deposited film structure, film bonding, and film properties are determined by a host of variables, including the promotion of surface reactions by ion, electron, or photon bombardment. The fundamental chemistry of these surface reactions is poorly understood, and accurate rate expressions are particularly

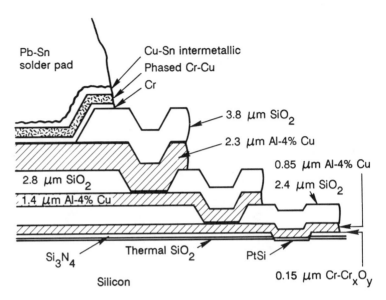

FIGURE 4.15 Cross-section of multilevel interconnections for advanced bipolar devices. Fourteen separate layers are laid down in the fabrication of interconnections such as the one shown. The precise orientation and composition of these layers are controlled by chemical process steps. Copyright 1982 by the International Business Machines Corporation. Reprinted with permission.

needed for electron-impact reactions (i.e., dissociation, ionization, and excitation), ion-ion reactions, neutral-neutral reactions, and ion-neutral reactions. The scale and scope of effort devoted in recent years to understanding catalytic processes need to be given to research on film deposition and plasma etching. Until we have a basic understanding of chemical reactions occurring at the surface and in the gas phase, it will be difficult to develop new etching systems.

Research in this area has had a demonstrable impact on recent innovations in plasma processing. Five years ago, it was well known that a fluorine-containing plasma etches silicon at a rate significantly greater than the rate for SiO_2, thus offering significant advantages for fabricating integrated circuits. However, well-controlled processes could not be developed that would perform in a production environment. The work of chemists and chemical engineers in elucidating the relevant chemical reactions and their kinetics was crucial to the identification of the important chemical species in the

etching process and their reaction pathways. In addition, this work led to the discovery that the organic polymer photoresist contributed to plasma chemistry and selectivity in important ways. This in turn led to new, improved plasma processes that are currently being used in production.

For magnetic media, mathematical models could enhance our fundamental understanding of the manufacturing processes used to make uniform high-purity magnetic particles. Models for the kinetics and mechanisms of reactions and an improved understanding of the thermodynamics of producing inorganic salts are required.

Modeling to describe the flows of viscous fluids could lead to better packaging of integrated circuits by assisting in the development of molding compounds and processes that will provide for lower thermal shrinkage stresses, lower permeability, and lower thermal conductivity. Such modeling could also contribute to the development of packaging materials and processes amenable to automation.

Engineering for Environmental Protection and Process Safety

Safety and environmental protection are extremely important concerns that present demanding intellectual challenges. The manufacture of materials and devices for information handling and storage involves substantial quantities of toxic, corrosive, or pyrophoric chemicals (e.g., hydrides and halides of arsenic, boron, phosphorus, and silicon; hydrocarbons and organic chlorides, some of which are suspected carcinogens; and inorganic acids). The expertise of chemical engineers in the safe handling and disposal of highly reactive materials is much needed in the electronics industry. Recent studies in California indicate that the semiconductor industry has an occupational illness rate three times that of general manufacturing industries. Nearly half of these illnesses involve systemic poisoning from exposure to toxic materials. Problems with groundwater contamination in Santa Clara County, California, have also raised concerns about how well the semiconductor industry is equipped to handle waste management and disposal. If the semiconductor and other advanced materials industries are to continue to prosper in the United States, it is important that the expertise of chemical engineers be applied to every aspect of chemical handling in manufacturing, from procurement through use to disposal.

IMPLICATIONS OF RESEARCH FRONTIERS

Industry has been the prime mover in advancing technology in electronic, photonic, and recording materials and devices. It will remain so for the foreseeable future. University research groups need to develop and maintain good communication with counterpart research groups in industry. Collaborative mechanisms are needed to promote academic-industrial coupling.

This coupling will become even more important as the electronics industry hires ever greater numbers of chemical engineers. Since

TABLE 4.4 Employment of Chemical Engineers in the Electronics Industry, 1977–1986[a]

Year	Number of Chemical Engineers
1977	700
1980	960
1983	1,648
1986	2,100

[a] Employment figures for Standard Industrial Classification code 367, "Electronic components and accessories."

SOURCE: National Science Foundation.[3]

1977, the number of chemical engineers employed by the industry has tripled (Table 4.4). Up to 25 percent of the recent graduating classes of several leading chemical engineering departments have been employed by the electronics industries. The increasing demand of these industries for chemical engineers is one factor to consider in planning for the support of the field. Any new mechanisms proposed must address this need.

In the electronics industry, a large number of relatively small firms play a key role in generating new process concepts and equipment. These firms face important research problems in fundamental science and engineering that would benefit markedly from the insights of academic chemical engineering researchers. Academic researchers should seek out and forge links to these small firms that stand at the crucial step between laboratory research and production processes. Potential mechanisms for accomplishing this are described in Chapter 10.

The current undergraduate curriculum in chemical engineering, although it provides an excellent conceptual base for graduates who move into the electronics industries, could be improved by the introduction of instructional material and example problems relevant to the challenges outlined in this chapter. This would not require the creation of new courses, but rather the provision of material to enrich existing ones. This theme is echoed, more broadly, in Chapter 10.

NOTES

1. National Research Council, National Materials Advisory Board. *State of the Art Reviews: Advanced Processing of Electronic Materials in the United States and Japan.* Washington, D.C.: National Academy Press, 1986.
2. National Academy of Sciences-National Academy of Engineering- Institute of Medicine, Committee on Science, Engineering, and Public Policy. "Research Briefing on High-Temperature Superconductivity," in *Research Briefings 1987.* Washington, D.C.: National Academy Press, 1987.
3. (a) National Science Foundation, Division of Science Resources Studies. *Employment of Scientists, Engineers, and Technicians in Manufacturing Industries: 1977* (NSF 80–306). Washington, D.C.: U.S. Government Printing Office, 1980.
   (b) National Science Foundation, Division of Science Resources Studies. *Scientists, Engineers, and Technicians in Manufacturing and Non-Manufacturing Industries: 1980–81* (NSF 83–324). Washington, D.C.: U.S. Government Printing Office, 1983.
   (c) National Science Foundation, Division of Science Resources Studies. *Scientists, Engineers, and Technicians in Manufacturing Industries: 1983* (NSF 85–328). Washington, D.C.: U.S. Government Printing Office, 1985.
   (d) Preliminary data from the 1986 survey of manufacturing industries provided by the NSF Division of Science Resources Studies.

SUGGESTED READING

M. Bohrer, J. Amelse, P. Narasimham, B. Tariyal, J. Turnipseed, R. Gill, W. Moebuis, and J. Bodeker. "A Process for Recovering Germanium from Effluents of Optical Fiber Manufacturing." *J. Lightwave Tech.*, LT-3 (3), 1984, 699.

T. Li, ed. *Optical Fiber Communications*, Vol. 1. Orlando, Fla.: Academic Press, 1984.

P. D. Maycock and E. N. Striewalt. *A Guide to the Photovoltaic Revolution: Sunlight to Electricity in One Step.* Emmaus, Pa.: Rodale Press, 1984.

R. H. Perry and A. A. Nishimura. "Magnetic Tape Production," in *Kirk-Othmer Encyclopedia of Chemical Technology*, 3rd ed., Vol. 14, p. 744. New York: Wiley-Interscience, 1979.

Solar Engineering Research Institute. *Basic Photovoltaic Principles and Methods* (FT-290–1448). Washington, D.C.: U.S. Government Printing Office, 1982.

S. M. Sze. *Semiconductor Devices: Physics and Technology.* New York: John Wiley & Sons, 1984.

W. Thomas, ed. *SPSE Handbook of Photographic Science and Engineering.* New York: Wiley-Interscience, 1973.

L. F. Thompson, C. G. Willson, and M. J. Bowden. *Introduction to Microlithography* (ACS Symposium Series No. 219). Washington, D.C.: American Chemical Society, 1983.

FIVE

Polymers, Ceramics, and Composites

Chemical engineers have long been involved with materials science and engineering. This involvement will increase as new materials are developed whose properties depend strongly on their microstructure and processing history. Chemical engineers will probe the nature of microstructure—how it is formed in materials and what factors are involved in controlling it. They will provide a new fusion between the traditionally separate areas of materials synthesis and materials processing. And they will bring new approaches to the problems of fabricating and repairing complex materials systems.

A FEW YEARS AGO, who would have dreamed that an aircraft could circumnavigate the earth without landing or refueling? Yet in 1986 the novel aircraft *Voyager* did just that (Figure 5.1). The secret of *Voyager's* long flight lies in advanced materials that did not exist a few years ago. Much of the airframe was constructed from strong, lightweight polymer-fiber composite sections assembled with durable, high-strength adhesives; the engine was lubricated with a synthetic multicomponent liquid designed to maintain lubricity for a long time under continuous operation. These special materials typify the advances being made by scientists and engineers to meet the demands of modern society.

The future of industries such as transportation, communications, electronics, and energy conversion hinges on new and improved materials and the processing technologies required to produce them. Recent years have seen rapid advances in our understanding of how to combine substances into materials with special, high-performance properties and how to best use these materials in sophisticated designs.

Chemical engineers have long been involved in materials science and engineering and will become increasingly important in the future. Their contributions will fall in two categories. For commodity materials, which are nonproprietary formulations with well-established chemical compositions and property standards, chemical engineers will help maintain U.S. competitiveness by creating and improving processes to make these chemicals as pure as possible and in high yields at the lowest possible investment and operating costs. For advanced materials, which are generally multicomponent, often proprietary, compositions designed to have very specific performance properties in specific uses, the competitive edge will come from chemical engineers who excel in controlling

FIGURE 5.1 The first airplane to circle the globe without refueling was *Voyager*, which accomplished this feat in 1986. This novel aircraft was made possible by high-performance lightweight materials and adhesives that were used in its construction. Chemical engineering research is crucial to the design of such new materials and their large-scale, efficient production. Copyright 1986 by Doug Shane Visions.

molecular conformation, microscopic and macroscopic structure, and methods of combining the components in a way that will maximize product performance.

Chapter 4 discussed chemical engineering challenges presented by materials and chemically processed devices for information storage and handling. In this chapter, five additional classes of materials are covered: polymers, polymer composites, advanced ceramics, ceramic composites, and composite liquids.

CHALLENGES TO CHEMICAL ENGINEERS

The revolution in materials science and engineering presents both opportunities and challenges to chemical engineers. With their basic background in chemistry, physics, and mathematics and their understanding of transport phenomena, thermodynamics, reaction engineering, and process design, chemical engineers can bring innovative solutions to the problems of modern materials technologies. But it is imperative that they depart from the traditional "think big" philosophy of the profession; to participate effectively in modern materials science and engineering they must learn to "think small." The crucial phenomena in making modern advanced materials occur at the molecular and microscale levels, and chemical engineers must understand and learn to control such phenomena if they are to engineer the new products and processes for making them. This crucial challenge is illustrated in the selected materials areas described in the following sections.

Polymers

The modern era of polymer science belongs to the chemical engineer. Over the years, polymer chemists have invented a wealth of novel macromolecules and polymers. Yet understanding how these molecules can be synthesized and processed to exhibit their maximum theoretical properties is still a frontier for research. Only recently has modern instrumentation been developed to help us understand the fundamental interactions of macromolecules with themselves, with particulate solids, with organic and inorganic fibers, and with other surfaces. Chemical engineers are using these tools to probe the microscale dynamics of macromolecules. Using the insight gained from these techniques, they are manipulating macromolecular interactions both to develop improved processes and to create new materials.

The power of chemical processing for controlling materials structure on the microscale is illustrated by the current generation of high-strength polymer fibers, some of which have strength-to-weight ratios an order of magnitude greater than steel. The best known example of these fibers, Kevlar™, is prepared by spinning an aramid polymer from an anisotropic phase (a liquid phase in which molecules are spontaneously oriented over microscopic dimensions). This spontaneous orientation is the result of both the processing conditions chosen and the highly rigid linear molecular structure of the aramid polymer. During spinning, the oriented regions in the liquid phase align with the fiber axis to give the resulting fiber high strength and rigidity. The concept of spinning fibers from anisotropic phases has been extended to both solutions and melts of newer polymers, such as

polybenzothiazole, as well as traditional polymers such as polyethylene. Ultrahigh-strength fibers of polyethylene have been prepared by gel spinning. The same concept, controlling the molecular orientation of polymers to produce high strength, is also being achieved through other processes, such as fiber-stretching carried out under precise conditions.

In addition to processes that result in materials with specific high-performance properties, chemical engineers continue to design new processes for the low-cost manufacture of polymers. The UNIPOL process for the manufacture of polyethylene is a good example of the contributions of engineering research to polymer processing. Polyethylene is probably the quintessential commodity polymer. It has been manufactured worldwide for decades, and current U.S. production exceeds 15 billion pounds per year. Considering the global capital investment in existing plants for making polyethylene, it could be argued that inventing a new process for its manufacture is a waste of time and money. Not so. Chemical engineers at Union Carbide designed a proprietary catalyst that allowed polyethylene to be made in a fluidized-bed, gas-phase reactor operating at low temperature and pressure (below 100°C and 21 Bar). The resulting process produces a polymer with exceptional uniformity and can precisely control the molecular weight and density of the product. The advantages of the process (including a low safety hazard from the mild operating conditions and minimal environmental impact since there are no liquid effluents and unreacted gases are recirculated) are such that, in 1986, UNIPOL process licensees had a combined capacity sufficient to supply 25 percent of the world's demand for polyethylene. This is remarkable market penetration for a new process technology for a mature commodity, particularly in light of the tremendous existing (and fully amortized) worldwide capacity for polyethylene. In 1985, Union Carbide and Shell Chemical successfully extended the UNIPOL process to the manufacture of polypropylene, another major polymer commodity. Interestingly, the first two licensees for the new polypropylene process were a Japanese chemical company and a Korean petroleum company.

Polymer Composites

Polymer composites consist of high-strength or high-modulus fibers embedded in and bonded to a continuous polymer matrix (Figure 5.2). These fibers may be short, long, or continuous. They may be randomly oriented so that they impart greater strength or stiffness in all directions to the composite (isotropic composites), or they may be oriented in a specific direction so that the high-performance characteristics of the composite are exhibited preferentially along one axis of the material (anisotropic composites). These latter fiber composites are based on the principle of one-dimensional microstructural reinforcement by disconnected, tension-bearing "cables" or "rods."

To achieve a material with improved properties (e.g., strength, stiffness, or toughness) in more than one dimension, composite laminates can be formed by bonding individual sheets of anisotropic composite in alternating orientations. Alternatively, two-dimensional reinforcement can be achieved in a single sheet by using fabrics of high-performance fibers that have been woven with enough bonding in the cross-overs that the reinforcing structure acts as a connected net or trusswork. One can imagine that an interdisciplinary collaboration between

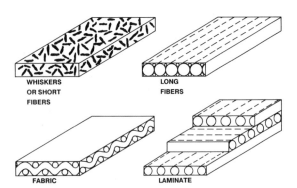

FIGURE 5.2 Fibers that are either very strong or very stiff can be used to reinforce polymers and ceramics. The resulting materials, known as composites, usually have one of the structures depicted in this figure. Clockwise from the upper left, reinforcement may be accomplished by embedding randomly oriented fibers, by orienting fibers along a particular axis, by assembling reinforced layers into laminates, or by embedding fabrics of reinforcing fibers in the material.

chemical engineers and textile engineers might lead to ways of selecting the warp, woof, and weave in fabrics of high-strength fibers to end up with trussworks for composites with highly tailored dimensional distributions of properties.

First-generation polymer composites (e.g., fiberglass) used thermosetting epoxy polymers reinforced with randomly oriented short glass fibers. The filled epoxy resin could be cured into a permanent shape in a mold to give lightweight, moderately strong shapes.

The current generation of composites is being made by hand laying woven glass fabric onto a mold or preform, impregnating it with resin, and curing to shape. Use of these composites was pioneered for certain types of military aircraft because the lighter airframes provided greater cruising range. Today, major components for aircraft and spacecraft are manufactured in this manner, as are an increasing number of automobile components. The current generation of composites are being used in automotive and truck parts such as body panels, hoods, trunk lids, ducts, drive shafts, and fuel tanks. In such applications, they exhibit a better strength-to-weight ratio than metals, as well as improved corrosion resistance. For example, a polymer composite automobile hood is slightly lighter than one of aluminum and more than twice as light as one of steel. The level of energy required to manufacture this hood is slightly lower than that required for steel and about 20 percent of that for aluminum; molding and tooling costs are lower and permit more rapid model changeover to accommodate new designs. Polymer composite hoods and trunk lids are commercial on the 1987 models of one major U.S. automobile line, and the early problems of higher manufacturing cost and of achieving adequate production have been largely overcome.

The mechanical strength exhibited by these composites is essentially that of the reinforcing glass fibers, although this is often compromised by structural defects. Engineering studies are yielding important information about how the properties of these structures are influenced by the nature of the glass-resin interface and by structural voids and similar defects and how microdefects can propagate into structural failure. These composites and the information gained from studying them have set the stage for the next generation of polymer composites, based on high-strength fibers such as the aramids.

Advanced Ceramics

For most people, the word "ceramics" conjures up the notion of things like china, pottery, tiles, and bricks. Advanced ceramics differ from these conventional ceramics by their composition, processing, and microstructure. For example:

• Conventional ceramics are made from natural raw materials such as clay or silica; advanced ceramics require extremely pure manmade starting materials such as silicon carbide, silicon nitride, zirconium oxide, or aluminum oxide and may also incorporate sophisticated additives to produce specific microstructures.

• Conventional ceramics initially take shape on a potter's wheel or by slip casting and are fired (sintered) in kilns; advanced ceramics are formed and sintered in more complex processes such as hot isostatic pressing.

• The microstructure of conventional ceramics contains flaws readily visible under optical microscopes; the microstructure of advanced ceramics is far more uniform and typically is examined for defects under electron microscopes capable of magnifications of 50,000 times or more.

Advanced ceramics have a wide range of application (Figure 5.3). In many cases, they do not constitute a final product in themselves, but are assembled into components critical to the successful performance of some other complex system. Commercial applications of advanced ceramics can be seen in cutting tools, engine nozzles, components of turbines and turbochargers, tiles for space vehicles, cylinders to store atomic and chemical waste, gas and oil drilling valves, motor plates and shields, and electrodes for corrosive liquids.

Because advanced ceramics provide key components to other technologies for major improvements in performance, their impact on the U.S. economy is much greater than is indicated by their sales figures. Ceramic components used

in turbines permit the construction of engines that operate at much higher temperatures than metallic engines, thus greatly increasing their thermodynamic efficiency and compactness. Ceramic liners and other ceramic components in diesel engines provide added benefits, such as the elimination of the need for water cooling and the prompter ignition of the fuel. An investment in wear-resistant ceramic cutting tools can be more than repaid by the decrease in downtime for sharpening or replacing a dulled or worn metallic tool.

Given these advantages, it is not surprising that market forecasts for advanced ceramics (including ceramic composites) are optimistic; in fact, sales in the year 2000 are predicted to be $20 billion. The market for advanced ceramics in heat engines is slated to grow by 40 percent per year to a total of $1 billion in 2000. The use of advanced ceramics is predicted to grow 16 percent per year over the next 5 years, and sales for automotive applications are forecast to increase from $53 million per year in 1986 to $6 billion per year by the end of the century.

Uniform microstructure is crucial to the superior performance of advanced ceramics. In a ceramic material, atoms are held in place by strong chemical bonds that are impervious to attack by corrosive materials or heat. At the same time, these bonds are not capable of much "give." When a ceramic material is subjected to mechanical stresses, these stresses concentrate at minute imperfections in the microstructure, initiating a crack. The stresses at the top of the crack exceed the threshold for breaking the adjacent atomic bonds, and the crack propagates throughout the material causing a catastrophic brittle failure of the ceramic body. The reliability of a ceramic component is directly related to the number and type of imperfections in its microstructure.

As the requirements for greater homogeneity in ceramics become more stringent, and the scale at which imperfections occur becomes smaller, the need for chemical processing of ceramics becomes more compelling. Traditional approaches to controlling ceramic microstructure, such as the grinding of powders, are reaching the limits of their utility for microstruc-

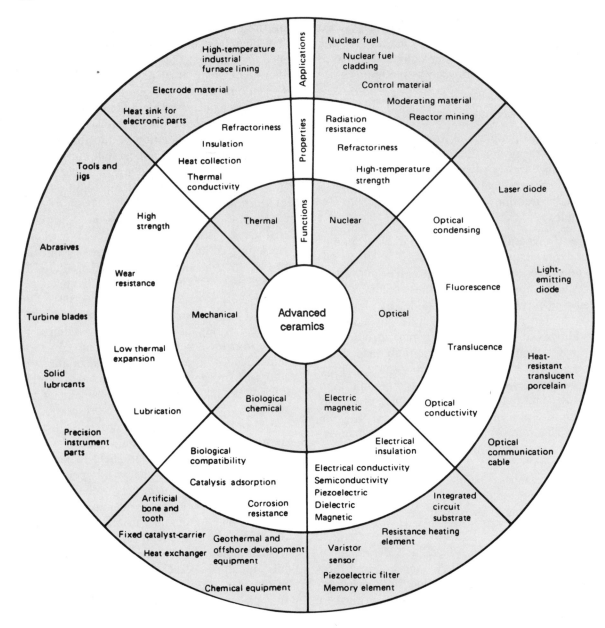

FIGURE 5.3 The myriad functions, properties, and applications of advanced ceramics. Reprinted from *High-Technology Ceramics in Japan*, National Materials Advisory Board, National Research Council, 1984.

tural control. Chemical engineers have an unparalleled opportunity to contribute their expertise in reaction engineering to problems that are in need of new analytical, synthetic, and processing tools. These include sol-gel processing and the use of chemical additives in ceramic processing.

Sol-Gel Processing

The use of sol-gel techniques to prepare ceramic powders has recently attracted much interest in academia and industry. Sol-gel techniques involve dissolving a ceramic precursor (e.g., tetramethyl orthosilicate) in a solvent and

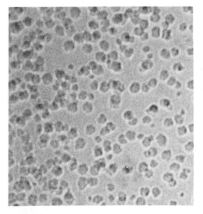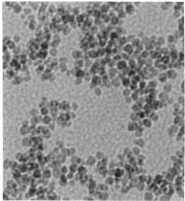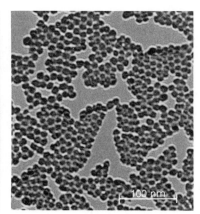

FIGURE 5.4 Stages in sol-gel processing are captured by a new electron microscopy technique. (1) Spherical particles tens of nanometers across can be seen in a colloidal silica sol. (2) Addition of a concentrated salt solution initiates gelation. (3) The gelled sample, after drying under the electron beam of the microscope, shows a highly porous structure. Courtesy, J. R. Bellare, J. K. Bailey, and M. L. Mecartney, University of Minnesota.

subjecting it to a carefully controlled chemical reaction, hydrolysis (Figure 5.4). When the hydrolysis products first appear as a separate phase, they are fine colloids consisting of small particles, some with radii as small as a few nanometers. This colloidal suspension (the sol) further reacts and polymerizes to form a porous high-molecular-weight solid (the gel) that contains the solvent as a highly dispersed fluid component in its internal network structure. Removal of the solvent leaves behind solids with a wide variety of macrostructures depending on the solvent and the way in which it was removed. These macrostructures can be sintered to convert them to dense ceramics.

Sol-gel techniques are of interest because they can be used to prepare powders with a narrow distribution of particle size. These small particles undergo sintering to high density at temperatures lower by several hundred degrees centigrade than those used in conventional ceramic processing. Sol-gel processes may also be used to prepare novel glasses and ceramics such as

• ceramics with novel microstructures and distributions of phases,
• amorphous powders and dried gels that can be processed without crystallization to fully

dense amorphous materials whose synthesis might not otherwise be possible,
• materials with controlled degrees of porosity and possibly tailored surfaces within pores, and
• ceramics with surfaces modified to alter their response to mechanical forces or to promote their adhesion to other materials.

Sol-gel processes also allow the manufacture of preforms that, upon sintering, collapse to a final product with the proper shape.

There are many unresolved problems in sol-gel processing, many of which revolve around the poorly characterized chemistry of the process. Understanding and controlling the polymerization reactions that produce the gel are key challenges, as are characterizing and optimizing both the removal of fluid from the gel and the subsequent sintering of the porous solid to a fully dense ceramic body. Solving these problems will make sol-gel processing the process of choice for the synthesis of a wide variety of ceramics, glasses, and coatings.

Chemical Additives in Ceramic Processing

Another area to which chemical engineers can contribute is the use of chemical additives to improve the properties of ceramic materials. For example, zirconium oxide can form a metastable state in ceramic bodies that is denser than its normal state. The incorporation of suitable chemical additives stabilizes the metastable state sufficiently to allow the fabrication of parts containing it. When a crack forms in such a ceramic part, the zirconium oxide region at the crack tip changes to the less dense form. The resulting expansion blunts the crack tip and stops its propagation (Figure 5.5). This strategy for using a chemical additive to improve ceramic resistance to cracking is called transformation toughening.

Ceramic Composites

Like polymer composites, ceramic composites consist of high-strength or high-modulus fibers embedded in a continuous matrix. Fibers may be in the form of "whiskers" of substances such as silicon carbide or aluminum oxide that are grown as single crystals and that therefore have fewer defects than the same substances in a bulk ceramic (Figure 5.6). Fibers in a ceramic composite serve to block crack propagation; a growing crack may be deflected to a fiber or might pull the fiber from the matrix. Both processes absorb energy, slowing the propagation of the crack. The strength, stiffness, and toughness of a ceramic composite is principally a function of the reinforcing fibers, but the matrix makes its own contribution to these properties. The ability of the composite material to conduct heat and current is strongly influenced by the conductivity of the matrix. The interaction between the fiber and the matrix is also important to the mechanical properties of the composite material and is mediated by the chemical compatibility between fiber and matrix at the fiber surface. A prerequisite for adhesion between these two materials is that the matrix,

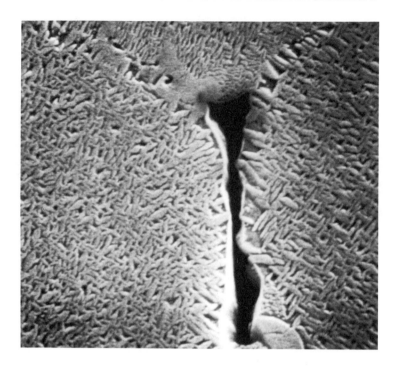

FIGURE 5.5 Zirconia ceramics can be made stronger and less brittle by using chemical additives to stabilize a more compact tetragonal structure that does not naturally occur at room temerature. When such a phase is subjected to stress, it can change phases, expanding to the monoclinic structure. This expansion fills any stress-initiated crack and prevents it from moving. This micrograph shows a zirconia ceramic composed of lozenge-shaped grains. A pore with grain boundaries radiating from its top and base dominates the picture. Courtesy, National Physical Laboratory (United Kingdom).

in its fluid form, be capable of wetting the fibers. Chemical bonding between the two components can then take place.

Ceramic matrix composites are produced by one of several methods. Short fibers and whiskers can be mixed with a ceramic powder before the body is sintered. Long fibers and yarns can be impregnated with a slurry of ceramic particles and, after drying, be sintered. Metals (e.g., aluminum, magnesium, and titanium) are frequently used as matrixes for ceramic composites as well. Ceramic metal-matrix composites are fabricated by infiltrating arrays of fibers with molten metal so that a chemical reaction between the fiber and the metal can take place in a thin layer surrounding the fiber.

As with advanced ceramics, chemical reactions play a crucial role in the fabrication of ceramic composites. Both defect-free ceramic fibers and optimal chemical bonds between fiber

FIGURE 5.6 This is a fractured sample of a ceramic composite (alumina with 30 volume-percent silicon carbide whiskers). The lighter regions of circular or cylindrical shape are randomly oriented whiskers protruding from the fractured surface. The rod-like depressions in the surface mark places where whiskers nearly parallel with the fracture were pulled out. Courtesy, Roy W. Rice, W. R. Grace and Company.

and matrix are required for these composites to exhibit the desired mechanical properties in use. Engineering these chemical reactions in reliable manufacturing processes requires the expertise of chemical engineers.

Composite Liquids

A final important class of composite materials is the composite liquids. Composite liquids are highly structured fluids based either on particles or droplets in suspension, surfactants, liquid crystalline phases, or other macromolecules. A number of composite liquids are essential to the needs of modern industry and society because they exhibit properties important to special end uses. Examples include lubricants, hydraulic traction fluids, cutting fluids, and oil-drilling muds. Paints, coatings, and adhesives may also be composite liquids. Indeed, composite liquids are valuable in any case where a well-designed liquid state is absolutely essential for proper delivery and action.

All composite liquids are produced by the

chemical processing industries, and chemical engineers face continuing challenges in tailoring their end-use properties. Some of these challenges are illustrated in the following examples:

• Motor lubricants are complex liquid composites in which components provide different performance characteristics. The basic component is a hydrocarbon oil with a fixed boiling range. It must have sufficient viscosity at engine operating temperatures to prevent the friction and wear of moving surfaces, but must be fluid enough below freezing temperatures for winter start-up. Viscosity modifiers are high-molecular-weight polymers that reduce the temperature coefficient of viscosity (viscosity index). Suspended colloidal particles of calcium or magnesium carbonate are added to neutralize engine acids and are stabilized by adsorbed polymers and surfactants to prevent coalescence. Solids dispersants are low-molecular-weight polymers with functional groups that pick up carbon particles generated in combustion and maintain them in suspension. At low temperatures, the waxes (straight-chain paraffin hydrocarbons) in the lubricant form long crystals to set up a solid gel. To prevent this, low-molecular-weight polymers, called pour point depressants, are added to co-crystallize with the wax; the resulting smaller crystals do not gel. Finally, there are antiwear additives and antioxidants to reduce engine wear and deposits. Lubricants with outstanding viscosity indexes enable an engine to start when the lubricant temperature is as low as −40°C and yet operate well when the lubricant temperature is as high as 200°C. Other additives allow broadening the temperature range further by providing increased thermal and oxidative stability. The use of synthetic base oils allows still broader ranges of operating temperatures, up to 500°C.

• Advanced adhesives are composite liquids that can be used, for example, to join aircraft parts, thus avoiding the use of some 30,000 rivets that are heavy, are labor-intensive to install, and pose quality-control problems. Adhesives research has not involved many chemical engineers, but the generic problems include surface science, polymer rheology and thermodynamics, and molecular modeling of materials near or at interfaces. The scientific and engineering skills needed are very similar to those needed for polymer composites and multicomponent polymer blends. The time-tested mechanical methods developed for joining metals are not satisfactory for composite and other advanced materials, and chemical engineers skilled in interfacial science are well qualified to contribute to this area.

• Another class of liquid composites is that of coating compositions used to deposit thin films on a substrate or other films; these composites have evolved from typical paints and varnishes into multilayer films in which each layer contributes specific properties to the ensemble. Such films may be paints used for sealing and decorative purposes, films used for printing or packaging purposes, or multilayer products used in recording tapes and photographic products. All are based on generic scientific principles that include many common elements from thermodynamics, polymer science, rheology, and fluid mechanics.

Liquid composites seldom behave as Newtonian fluids. These complex mixtures usually contain macromolecules, suspended particles, and surfactants. They are frequently multiphase, and changes in phase composition or formation of new liquid phases may occur over the range of operating conditions. Phase composition may be shifted by chemical reaction, by shear forces, or simply by changes in temperature or pressure. Liquid-liquid and liquid-solid equilibria are crucial. Detailed molecular understanding of the interactions among such components as surfactants, polymers, and particles is essential for the rational design of liquid composites. Much of this design is now accomplished by informed empiricism, which is useful for the incremental improvement of current products but inadequate for major changes and innovation.

INTERNATIONAL COMPETITION

The potential markets for the advanced materials discussed in this chapter are lucrative, and most nations that possess the technological infrastructure needed to invent, develop, and understand these materials are mounting major efforts to exploit these developing markets.

Polymers and Polymer Composites

The Panel on Advanced Materials of the NSF Japanese Technology Evaluation Program (JTECH) issued a report in 1986 assessing the status and direction of Japanese research and development efforts in several high-technology polymer areas.[1] The panel noted that the major Japanese chemical companies already manufacture most of the commercially available polymers and that "since 1970, there has been an increasing flow of upgraded technology from Japan to the United States." For engineering plastics and resins, the panel judged the United States to be ahead in basic research (although the lead is diminishing), on a par with Japan in advanced development, and behind Japan in product implementation.

The JTECH panel also compared the U.S./ Japanese position in high-strength/high-modulus polymer research and development. Its conclusions substantially agree with the following statements, drawn from a recent report of the NAS/NAE/IOM Committee on Science, Engineering, and Public Policy.[2] The United States has a strong position in the development of high-strength polymers, but comparable activity in this area exists in Japan. The Japanese Ministry of International Trade and Industry (MITI) has designated the development of a "third-generation" fiber as a government-subsidized project, beginning in 1983 and targeted for practical application by 1988. Because of the overwhelming importance of the load-bearing fibers in composites, the field is sensitive to breakthroughs in stronger fibers. Such a breakthrough could come from Japan; for example, a Japanese group was first to patent a process for making high-strength fibers from poly(ethylene terephthalate).

Much of the technology used for manufacturing carbon fibers in the United States is licensed from Japanese companies. The high level of Japanese carbon-fiber technology suggests that Japanese companies may produce many of the expected future advances in these materials.

The United States currently has a strong position in composite manufacturing and processing technology and leads the world in developing major applications in aircraft, sporting goods, and automotive components. There is growing overseas activity in composites technology, however, particularly among European aircraft companies.

Ceramics and Ceramic Composites

Japan is the United States' chief competitor in ceramics. There is a widespread but false perception that Japan leads the United States in ceramics in general. Nevertheless, it is clear that the Japanese effort in ceramics is comprehensive and long term and that several Japanese companies lead their U.S. industrial counterparts in specific technologies. There is tremendous enthusiasm in Japan for the potential of ceramics, and a recent report of a U.S. visiting team to Japan[3] reached the following conclusions:

• The Japanese are committed to vigorously developing and dominating the field of advanced ceramics. They have put in place a well-integrated national effort primarily based on government-industry interactions.
• Industrial management commitment to long-term research and development appears to be more solid in Japan than in the United States.
• Japanese research is focusing increasingly on fundamental research issues. Government-inspired basic research programs are being put in place, and Japan's position as a net consumer of basic ceramics research may change in the coming years.

Given the potential future importance of ceramics in areas as diverse as electronics (see Chapter 4), machine tools, heat engines, and superconductors (see Chapter 4), the United States can ill afford to surrender technical leadership to its competitors. The dominant trend in the field is toward materials with finer microstructures, fewer defects, and better interactions at interfaces (particularly in composites). Chemical processes provide important tools to capture the promise of ceramics for the benefit of our society and to maintain our international competitive position in technology.

Composite Liquids

The field of composite liquids has not received much attention outside the industries associated with specific liquid products (e.g., the petroleum industry). In areas such as lubrication, the United States has clear technological leadership. The situation is less clear for liquid crystals and adhesives, where there is greater competition from Europe and Japan.

INTELLECTUAL FRONTIERS

A wide variety of chemical engineering research frontiers involve advanced materials and belong in the mainstream of academic chemical engineering departments. The following list of frontiers ranges from the molecular level to the systems level.

Microscale Structures and Processes

The study of materials has traditionally centered on the influence of molecular composition and microstructure on mechanical, electrical, optical, and chemical properties. At the molecular level are a variety of research frontiers that can profitably draw chemical engineers into close collaboration with physical and theoretical chemists. They include the following research areas.

New Concepts in Molecular Design of Composite Materials

The toughest challenge and the greatest opportunity in chemical engineering for high-performance materials lie in the development of wholly new designs for composite solids. Such materials are typified by composites reinforced by three-dimensional networks and trussworks—microstructures that are multiply connected and that interpenetrate the multiply connected matrix in which they are embedded. In such materials, both reinforcement and matrix are continuous in three dimensions; the composite is bicontinuous. Geometric prototypes of such structures are found in certain liquid crystals and colloidal gels and in bones and shells. The challenge is to break away from today's technology, to go beyond today's research in two-dimensional fabric-like reinforcement, and to determine how to create truly three-dimensional microstructures and design, construct, and process them so as to control the properties of the composite.

A potentially promising area is "molecular composites," in which the fiber and its surrounding matrix have the same composition and differ only in molecular structure or morphology. This might involve forming the composite from very stiff, linear polymer molecules, some of which are aligned during the forming step as reinforcing crystallites in the amorphous regions—the matrix. An analogous ceramic composite may be envisioned. There are difficult engineering problems to be solved in learning how to control the orientation of the crystalline regions and the ratio between crystalline and amorphous regions in the material.

The Role of Interfaces in Materials Chemistry

Two general problems relate to the role of interfaces in advanced materials. The first is simply that we do not have the theory or the computational or experimental ability to understand the interatomic and microscopic interactions at the interfaces between components of an advanced material, on which its properties are critically dependent. There is a general need for research on processes at interfaces and on the structure-property-performance relationships of interfaces.

The second problem relates to the role that interfaces play in mediating chemical reactions in the synthesis of composite materials. This problem has three parts, which are illustrated here for polymeric composites.

• First, in composites with high fiber concentrations, there is little matrix in the system that is not near a fiber surface. Inasmuch as polymerization processes are influenced by the diffusion of free radicals from initiators and from reactive sites, and because free radicals can be deactivated when they are intercepted at solid boundaries, the high interfacial area of a prepolymerized composite represents a radically different environment from a conventional bulk polymerization reactor, where solid boundaries are few and very distant from the regions in which most of the polymerization takes place. The polymer molecular weight distribution and cross-link density produced under such diffusion-controlled conditions will differ appreciably from those in bulk polymerizations.

• Second, the molecular orientation of the fiber and the prepolymer matrix is important. The rate of crystal nucleation at the fiber-matrix interface depends on the orientation of matrix molecules just prior to their change of phase from liquid to solid. Thus, surface-nucleated morphologies are likely to dominate the matrix structure.

• Third, the ultimate mechanical properties of a composite will be strongly influenced by the degree to which the matrix wets the fiber surface and by the degree of adhesion between the two after curing. Both phenomena depend on intimate details of the surface science of the two phases, about which little is known.

Molecular modeling techniques, augmented by careful measurements of the structure of the interfacial regions, hold promise for elucidating details of these three aspects of interfacial control of matrix polymerization.

Understanding the Molecular Behavior of Complex Liquids

Basic understanding of the liquid state of matter still lags behind that of the solid and gaseous states. Our knowledge of interactions in multicomponent liquids containing macromolecules and suspended solids is extremely limited. Thus, the study of complex liquids, including polymer solutions, sols, gels, and composite liquids, is a significant challenge for chemical engineers.

The ability to predict liquid-liquid and liquid-solid equilibria in complex systems is still rather undeveloped, in part because of the lack of systematic and molecularly interpreted experimental information. Considerable research has been conducted on the behavior of liquids near their critical points, on lower critical solution phenomena, on spinodal decomposition, and on related dynamics such as the growth and morphology of new phases, but generalized correlations and connections of theory to practice are few.

Molecularly motivated empiricisms, such as the solubility parameter concept, have been valuable in dealing with mixtures of weakly interacting small molecules where surface forces are small. However, they are completely inadequate for mixtures that involve macromolecules, associating entities like surfactants, and rod-like or plate-like species that can form ordered phases. New theories and models are needed to describe and understand these systems. This is an active research area where advances could lead to better understanding of the dynamics of polymers and colloids in solution, the rheological and mechanical properties of these solutions, and, more generally, the fluid mechanics of non-Newtonian liquids.

Chemical Dynamics and Modeling of Molecular Processes

Chemical dynamics and modeling were identified as important research frontiers in Chapter 4. They are critically important to the materials discussed in this chapter as well. At the molecular scale, important areas of investigation include studies of statistical mechanics, molecular and particle dynamics, dependence of molecular motion on intermolecular and interfacial forces, and kinetics of chemical processes and phase changes.

Mechanistic studies are particularly needed for the hydrolysis and polymerization reactions that occur in sol-gel processing. Currently, little is known about these reactions, even in simple systems. A short list of needs includes such rudimentary data as the kinetics of hydrolysis and polymerization of single alkoxide sol-gel systems and identification of the species present at various stages of gel polymerization. A study of the kinetics of hydrolysis and polymerization of double alkoxide sol-gel systems might lead to the production of more homogeneous ceramics by sol-gel routes. Another major area

for exploration is the chemistry of sol-gel systems that might lead to nonoxide ceramics.

The Intimate Connection Between Materials Synthesis and Processing

Materials synthesis and materials processing have classically been thought of as separate activities, and in the days of simple, homogeneous materials, they were. But today's complex materials are bringing these two areas closer together in research and in practice. Four outstanding intellectual challenges demonstrating this connection are described in this section.

Processing of Complex Liquids

Complex liquids are ubiquitous in materials manufacture. In some cases, they are formed and must be handled at intermediate steps in the manufacture of materials (e.g., sols and gels in the making of ceramics, mixtures of monomer and polymer in reactive processing of polymers). In other cases (e.g., composite liquids), they are the actual products. Understanding the properties of complex fluids and the implications of fluid properties for the design of materials processes or end uses presents a formidable intellectual challenge.

Complex liquids seldom behave as classical Newtonian fluids; thus, analysis of their behavior requires a thorough understanding of non-Newtonian rheology. The importance of this knowledge is illustrated by the following two examples:

• The problem of processing complex liquids while they are undergoing rapid polymerization is an important challenge in reactive polymer processing (e.g., reactive injection molding and reactive extrusion). In these processes, the viscosity of a reaction mixture, as it proceeds from a feed of monomers to a polymer melt product, may change by 7 decades or more in magnitude. Fluid mixtures flowing into a mold of complicated geometry may exhibit large temperature gradients from the highly exothermic chemical reactions taking place and significant spatial variations in viscosity and molecular weight distribution.

• Rheology is especially important to the understanding of composite liquids in their many applications as products (e.g., lubricants, surface coating agents, and additives for enhanced oil recovery and drag reduction). Such liquids usually contain polymers, and their behavior is frequently viscoelastic under use conditions. While data on linear response and relatively mild shear flows are available for nonassociating polymer solutions in the relevant ranges of molecular size and concentration, far fewer data are available on liquid systems that contain particles or micelles, particularly those in which there are strong interparticle interactions. Knowledge of the fluid mechanics of ordered liquids is similarly sparse. Information on the response to rapid shear flows and extensional flows, even in simple polymer solutions, is very limited. Thus, we are far from having dependable equations from which models of such fluids could be developed and farther still from a generalized molecular understanding of the structure-property relations of these fluids and from extrapolations of the flow patterns and stress distributions in such fluids in geometries close to those in which they are used. For example, there is significant divergence between theoretical prediction and empirical observation of the flow of lubricants in journal bearings.

Even if satisfactory equations of state and constitutive equations can be developed for complex fluids, large-scale computation will still be required to predict flow fields and stress distributions in complex fluids in vessels with complicated geometries. A major obstacle is that even simple equations of state that have been proposed for fluids do not always converge to a solution. It is not known whether this difficulty stems from the oversimplified nature of the equations, from problems with numerical mathematics, or from the absence of a laminar steady-state solution to the equations.

Processing of Powders

One route to better ceramic powders, sol-gel processing, has already been described in this chapter. There are, however, many other possible routes to improved ceramic powders. These routes include refinements of older processes, such as precipitation and thermal decomposition, as well as newer processes, such as plasma processing and chemical vapor deposition. The nucleation and particle-growth processes in such systems need to be described quantitatively to enable better process development and scale-up. Chemical engineering frontiers include the development of new chemical processes for producing ceramic raw materials, such as submicron, spherical, uniform powders, and high-strength fibers and whiskers.

Chemical engineers could also work to devise processes to improve the flow characteristics of powders after they are formed. Such research would help control agglomeration of particles in subsequent processing steps as well as facilitate the production of compacted ceramic preforms. For example, gas-solid chemical reactions might be used to tailor the chemical composition of powders. As another example, better methods of compounding powders with binders might be achieved by processes that mix powders with suitable binders in a liquid and then spray dry the resulting suspension.

Powder processing is also one element in the engineering of grain boundaries in large, complex parts. Such engineering would allow sintering ceramics to full density without degrading oxidation resistance and long-term strength.

Processing of Polymers

Other important research challenges confront chemical engineers in the area of polymer processing. One concerns the interactions of polymers with their environment. For example, contacting a glassy polymer with a solvent or swelling agent may lead to unusual diffusion characteristics in the polymer, stress formation, crazing, or cracking. Such phenomena are poorly understood because glassy polymers may exhibit complex viscoelastic behavior in the presence of a liquid or during their second-order (glass) transitions. The study of diffusion in glassy polymers is a virgin research area for chemical engineers. A better understanding of polymer-solvent interactions could have important payoffs in the development of positive resists for microcircuit manufacture (see Chapter 4), because the dissolution characteristics of polymeric resists are crucial to their application and removal during microlithography.

The focus of research on engineering thermoplastics with enhanced mechanical, thermal, electrical, and chemical properties has shifted away from synthesizing novel polymers toward combining existing polymers. Multicomponent polymer blends pose interesting and challenging new problems for chemical engineers. Many multicomponent polymeric melts are homogeneous at processing temperatures but separate during cooling. Judicious choice of stress levels during cooling and of the cooling rate can effect changes in the structure and morphology of the end product and hence in its properties.

The fluid prepolymer in which the load-bearing fibers of a polymer composite are placed undergoes further polymerization and crosslinking during the thermal curing of the composite. The chemical reactions that occur during curing are exothermic and are difficult to control. Some regions in the composite material react adiabatically while others lose heat by conduction to their surroundings. The resulting point-to-point variations in polymer matrix molecular weight and cross-link density result in changes in the composite's properties and quality. We need to better understand and control these variations in well-characterized processes and to deduce how to change the geometry of the finished object or the distribution of fibers within it to compensate for the variations in polymer structure that might inherently arise during processing.

Process Design and Control

Because processing conditions and history have such an important influence on the conformation and properties of materials, there is a need to develop models and systems for the measurement and control of materials manufacturing processes so that processes can be better designed, more precisely controlled, and automated. Opportunities for chemical engineers in process design and control, including advanced mathematical modeling of polymer processing, are explored in depth in Chapter 8.

It is particularly important to study process phenomena under dynamic (rather than static) conditions. Most current analytical techniques are designed to determine the initial and final states of a material or process. Instruments must be designed for the analysis of materials processing in real time, so that the crucial chemical reactions in materials synthesis and processing can be monitored as they occur. Recent advances in nuclear magnetic resonance and laser probes indicate valuable lines of development for new techniques and comparable instrumentation for the study of interfaces, complex liquids, microstructures, and hierarchical assemblies of materials. Instrumentation needs for the study of microstructured materials are discussed in Chapter 9.

Fabrication and Repair of Materials Systems

Advanced materials systems based on polymers, ceramics, and composites are constructed by assembling components to create structures whose properties and performance are determined by the form, orientation, and complexity of the composite structure. The properties of these assemblages are determined not by the sum of weighted averages of the components but rather by synergistic effects in interconnected phases. For this reason, the study of fabrication of hierarchical assemblages of materials, as well as the study of mechanisms for repairing defects in assembled structures, must be supported by fundamental research.

Designing Systems from the Molecules on Up

Successful systems design and fabrication depend on understanding the connections between microscale phenomena and macroscale behavior of materials. For example, with sufficient insight into intermolecular interactions, appropriate models, and the computational power of supercomputers, it may be possible to predict changes in macromolecular configurations when loads are imposed on polymers or changes in the properties of a material as a result of branching or cross-linking the material's macromolecular structure.

A related problem in composites is the need to design optimal fiber orientations for a composite part given the set of stress vectors and levels to which the part will be subjected. These design considerations would be useful in designing airframe components such as parts for the tail, wing, or fuselage. A similar problem is assessment of the performance penalties that might result from imperfections in manufacture.

Solutions to these problems lie in the realm of computer-aided design and manufacturing (CAD/CAM). This area of technology is being developed rapidly by mechanical engineers, but the problems encountered include many that are logical extensions of polymer process engineering. Interdisciplinary collaborations between mechanical and chemical engineers should be fostered for problems where chemical expertise would be valuable. Just as chemical engineers of a previous era contributed extensively to the knowledge of heat and mass transfer by collaborating with mechanical engineers, so are they now well positioned to contribute to composites CAD/CAM and to the education of students who may one day use and oversee these processes in industry.

Chemical Processing in the Fabrication of Materials Systems

One might imagine that the fabrication of materials systems involving polymers, ceramics, and composites would be principally a concern of mechanically oriented materials engineers. This is not true. For example, the mechanical attachment of composites to other materials (e.g., metal parts) by drilling holes in the composite and attaching mechanical fasteners can alter and degrade the performance of the composite. In a number of situations, joining and fabrication processes involving chemical reactions with the material will be needed in systems fabrication.

Fundamental research to support materials assembly and fabrication probably centers on the science and technology of adhesion, although research on mechanical assembly driven by chemical action, such as the self-assembly

of large molecules or particles, also holds promise for solving some fabrication problems.

Detection and Repair of Flaws in Materials Systems

A central problem in complex materials systems of any kind involves testing to detect flaws, analysis to predict their effect on remaining service life of the system, and repair strategies to overcome them. For the structural materials discussed in this chapter, these problems are uncharted territory in need of exploration by chemical engineers.

There is a general need for nondestructive test methods capable of determining whether the manufacturing process for a polymer, ceramic, or composite has achieved the desired microstructure. Chemical engineers can profitably contribute to interdisciplinary efforts to develop such test methods. For example, one type of defect in a composite arises because the placement of the fibers is different from what was intended. This may reflect perturbations in the filament-winding operations on a mandrel or fiber movement during curing in response to differential stresses. The processing expertise of chemical engineers could be useful in developing instrumentation to detect such flaws during the manufacturing process so that automatic control and correction of the process can be invoked to avoid or compensate for flaws.

There is also a need for methods to predict the effects of flaws and the remaining service life of a flawed or degraded part in use. For example, because of limited basic knowledge about composites, structures based on them are now overdesigned for considerably greater margins for error than those required for metal structures, thereby losing some of the inherent superiority of composites.

Finally, attempts to repair composite structures will become increasingly common in future years as the use of composites spreads. At this point, a fundamental repair science for composites is completely lacking. Since such strategies are likely to depend heavily on chemical reactions to heal breaks and flaws, chemical engineers should be at the forefront of this emerging field.

IMPLICATIONS OF RESEARCH FRONTIERS

Chemical engineers are already equipped to pursue the frontiers outlined in this chapter. The core undergraduate curriculum provides both a science base and an engineering knowledge base for approaching problems in materials phenomena and processing. At the same time, undergraduate chemical engineering students would benefit from a broader exposure to problems in materials science and engineering. What is needed is a better integration of such problems into the curriculum at all levels. This is best achieved by developing better instructional material and example problems for existing courses in thermodynamics, transport, and reaction engineering. What is *not* needed is a proliferation of general, encyclopedic materials courses for undergraduates.

All the scientific and engineering disciplines involved in materials research are in need of better instrumentation and facilities. Suitable equipment for chemical engineers interested in materials questions might include the following:

- solid-state NMR spectrometry;
- spin-echo NMR spectrometry;
- Raman spectroscopy;
- secondary ion mass spectrometry;
- X-ray photoelectron spectroscopy;
- laser light scattering;
- advanced dynamic rheometers;
- computer-controlled, fully equipped polymerization reactors;
- directional irradiation devices; and
- dynamic mechanical property measurement equipment.

Special efforts to help academic institutions acquire these instruments are needed. Future chemical engineers will be required to understand the design and operation of sophisticated equipment in the analysis of materials properties. An early exposure to these techniques is highly desirable, and is probably indispensable to quality research at the graduate level.

The Materials Research Laboratories (MRLs), sponsored by the NSF, have been one mechanism for providing instrumentation and facilities support to small groups of principal investigators with interests in materials. There is a perception in the chemical engineering community that MRLs are more physics directed, and probably not open to significant participation by chemical engineers. One way of addressing this problem would be for NSF to target more funds to the Division of Materials Research with an emphasis on interdisciplinary and process studies. Another way might be to develop mechanisms intermediate between MRLs and ERCs that would promote engineering research on materials.

A final goal for improving the chemical engineering contribution to materials research would be to develop focused continuing education programs to help qualified chemical engineers move aggressively into materials-related areas. Such courses might take a number of forms. The AIChE might take the lead in sponsoring short courses within the context of its existing continuing education program. Universities might provide complementary, more intense exposure to the problems and opportunities in materials research by initiating special workshops, masters degree programs, or sabbaticals for industrial researchers.

NOTES

1. *JTECH Panel Report on Advanced Materials in Japan.* La Jolla, Calif.: Science Applications International Corp., 1986.
2. National Academy of Sciences-National Academy of Engineering-Institute of Medicine, Committee on Science, Engineering, and Public Policy. "Report of the Research Briefing Panel on High-Performance Polymer Composites," in *Research Briefings 1984*. Washington, D.C.: National Academy Press, 1984.
3. National Research Council, National Materials Advisory Board. *High-Technology Ceramics in Japan* (NMAB-418). Washington, D.C.: National Academy Press, 1984.

SUGGESTED READING

C. G. Gogos, Z. Tadmor, D. M. Kalyon, P. Hold, and J. A. Biesenberger. "Polymer Processing: An Overview." *Chem. Eng. Prog.*, 83 (6), June 1987, 49.

National Research Council, Engineering Research Board. "Materials Systems Research in the United

States," in *Directions in Engineering Research.* Washington, D.C.: National Academy Press, 1987.

D. R. Uhlmann, B. J. J. Zelinski, and G. E. Wnek. "The Ceramist as Chemist—Opportunities for New Materials." *Mat. Res. Soc. Symp. Proc.*, 32, 1984, 59.

U.S. Congress, Office of Technology Assessment. *New Structural Materials Technologies: Opportunities for the Use of Advanced Ceramics and Composites—A Technical Memorandum* (OTA-TM-E-32). Washington, D.C.: U.S. Government Printing Office, 1986.

SIX

Processing of Energy and Natural Resources

Energy, minerals, and metals are three basic building blocks of our technological society. Chemical engineering has long been involved in the technologies used to convert natural resources into energy and useful products. The expertise of chemical engineers will be needed more than ever if we are to make progress on problems such as enhanced oil recovery, shale oil production, coal conversion, electrochemical energy storage, solar power, and turning waste into a useful source of energy and metals. Important intellectual challenges await them in in-situ processing, processing solids, developing better separations, finding better materials for use in energy and mineral applications, and advancing the knowledge base for process design and scale-up. Implications of these challenges for chemical engineers are discussed at the end of the chapter.

"THE MORAL EQUIVALENT OF WAR." This summons to respond to the energy crisis, uttered by President Jimmy Carter less than 10 years ago, today seems so remote that it might be the subject of a good trivia question. The mid-1980s are witness to a vast demobilization of resources and personnel once committed to energy research. Armies of researchers have been redirected, retired, or laid off. Pilot and demonstration facilities have been mothballed. Spending on alternative fuels research has dwindled. The cumulative effect of this rapid negative change is to discourage academic investigators and young graduates from seriously considering the formidable energy research problems that remain unsolved.

Yet there is nothing trivial about energy. The cost of energy over the last decade has had a significant influence on the rate of inflation in the United States. The energy processing industries constitute one of the country's largest industrial segments. In 1985, shipments of petroleum and coal products amounted to $194 billion,[1] and the value of natural gas produced in the United States exceeded $42 billion.[2] The availability of secure fuel supplies is vital to national defense. Cheap and abundant energy supplies are just as important to national economic competitiveness in peacetime. No industrial segment has a greater impact on national well-being, jobs, defense, and economic competitiveness. No industrial segment is more dependent on chemical engineers for its health and progress. Research, development, and commercial operations in the energy processing industries all draw heavily on the knowledge and techniques of chemical engineers.

There is nothing trivial about our nation's primary metals industries either—in 1985 they accounted for $126 billion in shipments.[1] This industrial sector and the energy industries have a number of characteristics in common. Both face an uncertain future shaped by the declining quality of domestic raw materials reserves and substantially higher quality reserves in countries that need to obtain foreign exchange through commodity sales, regardless of market price or profit considerations. The United States depends heavily on foreign imports for both energy and metals resources (Figure 6.1). The natural

resources industries have experienced cutbacks in support for industrial and academic research that are similar to those in the energy industries.

Energy processing and natural resources processing share numerous fundamental technical problems, many of which fall squarely in the domain of the chemical engineer. The current retrenchment in research in both fields does not imply that the problems have largely been solved. Indeed, the problems are as challenging as ever, and they are not going to be solved overnight. Now is the time to conduct the fundamental research needed for their solution.

How do we take a future-oriented approach to research on energy and metals? What criteria do we use to set research priorities? Short-term projections of prices and availability of resources are poor guides to a national policy for research. It is virtually impossible to predict the course that energy prices will take over the next few years or to prognosticate political events that might affect the supply of key minerals to the United States. The most anyone can say is that oil prices will rise and that a real threat exists to the stability of our supply of several key minerals.

What is predictable is that the cost of producing oil and minerals in the United States will be influenced by the rate of depletion of recoverable domestic sources. This rate will in turn be influenced by the economics of recovering usable materials from these domestic sources. It is certain that there will be a need for engineers to develop and manage technologies that can slow the depletion rate by permitting recovery of a greater fraction of the resources that are there.

The United States is the recognized world leader in energy and minerals technology; chemical engineering has been and must continue to be the key discipline in maintaining that position. The primary thrust of chemical engineering research in these areas should be to provide the basic knowledge that would give the nation the capability to react to future shifts in the prices and availability of energy and minerals. This capability can be developed by pursuing fundamental research in the priority areas laid out in this chapter. Our ultimate objectives should be to extract and process current resources

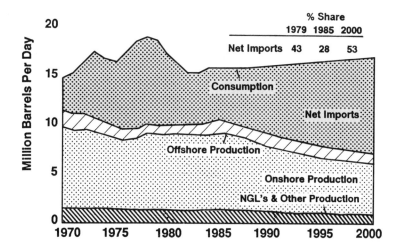

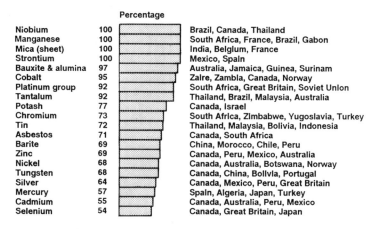

FIGURE 6.1 The United States is becoming ever more dependent on foreign sources of oil and minerals. The top graph displays trends in U.S. production and consumption of petroleum feedstocks from 1970 to 2000. It shows the growing contribution of imported oil to U.S. consumption, a contribution that is projected to increase rapidly in the 1990s. The bottom table shows that the United States depended in 1985 on foreign suppliers for 20 minerals and metals, some of which are critical to national security. Courtesy, Chevron Oil Company (top) and the U.S. Bureau of Mines (bottom).

more efficiently and to develop ways of exploiting alternative resources.

TECHNOLOGIES FOR EXPLOITING ENERGY SOURCES

In the early part of this century, technologies were developed for exploiting apparently limitless reservoirs of gaseous and liquid fossil fuels. Because these fuels are easier to handle and cleaner than coal, they made great inroads on the use of coal as a primary energy source. However, in recent years, it has become apparent that easily accessible, high-quality reservoirs of gaseous and liquid fossil fuels are not inexhaustible. Moreover, the demand for energy continues to increase as the population increases and labor-sparing machines are developed.

It is a basic premise that this challenge can best be met by developing technologies for more efficient use of existing resources and for utilizing previously untapped sources of energy. A number of approaches to these goals appear to hold promise for success in the next few decades.

Enhanced Oil Recovery

Technologies for oil production can be divided into three classes: primary recovery, secondary recovery, and enhanced oil recovery. In primary recovery, the oil and gas flow naturally through the reservoir rock to the production well, impelled by subterranean pressure. For typical light oils, only 15–20 percent of the oil in the formation is extracted in primary recovery. Secondary recovery processes extend primary recovery by injecting water or gas to maintain reservoir pressure as the oil is removed. These processes are well established and generally recover an additional 15–20 percent of the original oil. Thus, for conventional medium and light crude oils, which have relatively low viscosity at reservoir conditions, about one-third of the original oil can be recovered by primary and secondary methods.

Enhanced oil recovery (EOR) processes (also called tertiary recovery processes) are used to recover a portion of the remaining two-thirds of the original oil. Adverse reservoir properties and conditions limit both the applicability of EOR processes and the extent of recovery from those reservoirs to which the processes can be applied effectively. In addition, there are entire oil deposits so viscous—"ultraheavy" crudes, bitumens, and tars—that primary recovery is not possible and secondary recovery processes are generally ineffective. Such deposits in the United States approach the potential of conventional oil reserves, and on a worldwide basis they are several-fold greater than total reserves of the lower viscosity conventional crude oils. Much of the total resource of these extremely viscous oils is in Canada and Venezuela.

The ability to recover as much as possible of the remaining two-thirds of conventional oils in known formations and to utilize ultraheavy crude deposits will become increasingly important as U.S., and ultimately worldwide, reserves of conventional crude oils are depleted.

The three classes of EOR technologies that have been studied extensively are thermal recovery, miscible flooding, and chemical flooding. For each of these methods, the following two basic problems must be overcome if we are to recover a significant part of the remaining oil.

• Oil deposits are found in porous sedimentary rocks with limited pathways (permeability) for flow of oil through the reservoir formation to a producing well. The first problem is to achieve microscopic recovery efficiency—to displace the oil from the rock matrix and cause it to flow through the formation along a specific pathway. Figure 6.2 depicts an oil droplet in porous sand. Because of interfacial tension, the oil cannot be moved by water pumped into the formation. If the interfacial tension of the oil is lowered—whether by increased temperature, by an oil-miscible sweeping fluid, or by chemical additives—the oil can move from its original location through the porous sand.

• The second problem is to achieve macroscopic sweep efficiency—recovery of oil from a significant fraction of the reservoir formation. This problem is generally more intractable than the first. The reservoir formation is not homogeneous in porosity or permeability, which causes

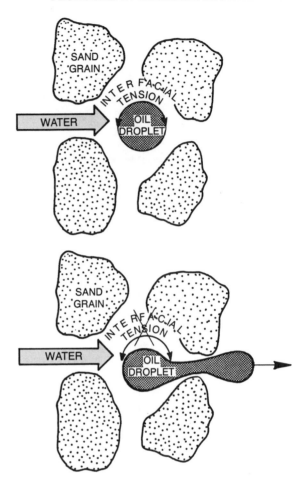

FIGURE 6.2 Interfacial tension imprisons residual oil in rock, preventing its displacement by water. Without interfacial tension, oil flows freely, leaving no residual portion in the rock. Courtesy, Amoco Production Company.

phenomena such as fingering and allows pockets of oil to be bypassed, dramatically lowering the sweep efficiency of all current EOR methods (Figure 6.3). A basic challenge for chemical engineers is how to detect and/or model the movement of oil, water, gas, and injected chemicals in the heterogeneous environment of the reservoir. This problem is discussed in detail in Chapter 8, which is devoted to computer-assisted process and control engineering.

Thermal recovery methods involve the use of steam and in-situ combustion. Thermal EOR processes add heat to the reservoir to reduce the viscosity of the oil or to vaporize it. In addition, these processes use steam or oil com-

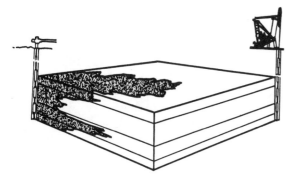

FIGURE 6.3 Oil can be recovered from reservoirs by pumping in steam, gas, or specialty chemicals. All methods face common problems posed by inhomogeneities in the rock containing the oil. Some major problems include poor vertical coverage, inefficient sweeping that bypasses pockets of oil, and severe channelling of fluids along fissures or highly permeable layers of rock. Chemical methods of compensating for these inhomogeneities would boost the yields and cut the operating costs of enhanced oil recovery. Courtesy, Amoco Production Company.

bustion products as a drive fluid to move oil to producing wells. Thermal processes are most often used in reservoirs of viscous oils and tars on which tests have established that primary production will be small and waterflooding will be largely ineffectual. Thus, thermal methods are usually used in place of, rather than after, secondary or primary methods.

The first EOR method to achieve widespread commercial acceptance, steam injection (Figure 6.4), has been used commercially in California for more than 20 years and now accounts for more than 80 percent of U.S. EOR production.

A less well developed thermal method, in-situ combustion, holds much promise but also poses a tremendous challenge to the theory and practice of chemical engineering. The in-situ combustion process involves many processes occurring simultaneously. Heat is generated within the reservoir by injecting air or oxygen to burn part of the reservoir oil (Figure 6.5). It is common practice to co-inject water with or above the oxidant to scavenge energy from hot rock lying behind the burn front. In-situ combustion often achieves considerably higher temperatures than steam flooding, and the heat not only physically and chemically reduces the oil viscosity but also partially vaporizes the oil,

which is driven forward by a combination of steam, hot water, and gas.

The earth itself is the reaction vessel and chemical plant. The complicated reaction chemistry and thermodynamics involve mixers, reactors, heat exchangers, separators, and fluid flow pathways that are a scrambled design by nature. Only the sketchiest of flowsheets can be drawn. The chemical reactor has complex and ill-defined geometry and must be operated in intrinsically transient modes by remote control. Overcoming these difficulties is a true frontier for chemical engineering research.

Another EOR approach to reducing the viscosity of oil in the reservoir is miscible flooding—the injection of fluids that mix with the oil under reservoir conditions. Such fluids include carbon dioxide, light hydrocarbons, and nitrogen. Supply and cost of carbon dioxide are often more favorable than for other injectants. Extensive research and field testing have established the technical viability of miscible flooding, and a number of commercial carbon dioxide miscible flooding projects are in operation.

Chemical EOR methods are based on the injection of chemicals to develop fluid or interfacial properties that favor oil production. The three most common of these methods are polymer flooding, alkaline flooding, and surfactant flooding.

Commercial implementation of polymer flooding and alkaline flooding is in progress, and there is confidence that research can make these processes more cost-effective and extend their applicability to a greater fraction of the known reservoirs. The research focus is on improving the thermal, chemical, and biological stability of polymers and making them more cost-effective. Research targets in alkaline flooding include better definition of the reactions of alkaline materials with the rock formation and the use of ancillary chemicals to improve performance. As with polymer flooding, a primary objective is to extend the applicability to more severe conditions.

Of the chemical EOR technologies, surfactant flooding is the most complex, the farthest from commercial feasibility, and the most challenging in terms of research needs, yet it has the greatest ultimate potential. It involves injecting surfac-

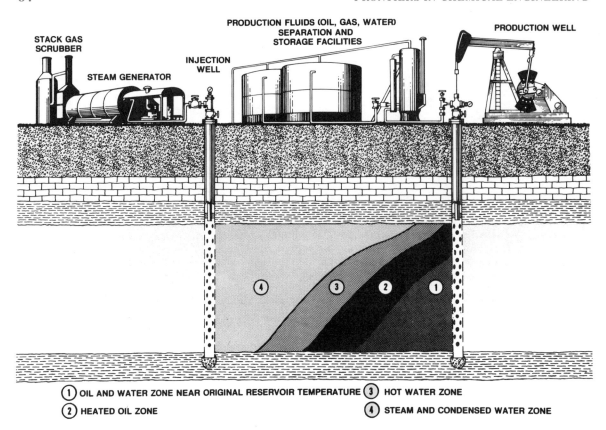

FIGURE 6.4 Steam flooding is one of two principal thermal methods for oil recovery and has been commercially applied since the early 1960s. A mixture of steam and hot water is continuously injected into the oil-bearing formation to displace mobilized oil to adjacent production wells. Reprinted with permission from *Enhanced Oil Recovery*. Copyright 1984 by the National Petroleum Council.

tants, such as sulfonated crude oil, to mobilize the oil for subsequent recovery by a waterflood. The most recent National Petroleum Council (NPC) study of EOR[3] estimates that chemical EOR technologies constitute more than 60 percent of the additional EOR potential for advanced technology and that surfactant flooding represents more than 90 percent of this potential. Research objectives are to extend surfactant flooding to more severe conditions and to make it more cost-effective.

The stakes in continued research and development of EOR technologies are enormous. They involve decreasing U.S. dependence on imported oil and extending the useful lifetime of the world's exhaustible supply of petroleum. The NPC study of EOR estimates that with currently implemented EOR technologies the

total ultimate EOR potential for the United States is 14.5 billion barrels (Figure 6.6), which is more than 50 percent of the U.S. total estimated future recovery by primary and secondary processes. The NPC study also estimates that successful development of projected advanced EOR technologies could make possible the recovery of 27.5 billion barrels of domestic oil (Figure 6.6). This is more than 10 years of U.S. production at current rates and could provide an important augmentation of domestic supplies well into the next century.

Shale Oil Production

Oil shales are a large, virtually untapped source of hydrocarbons. U.S. reserves represent several hundred billion barrels of oil and

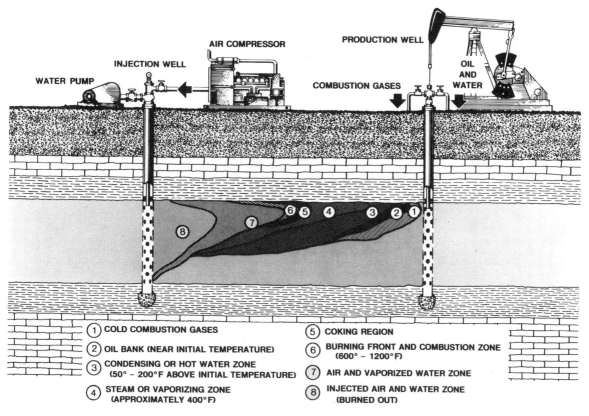

WATER PUMP INJECTION WELL AIR COMPRESSOR PRODUCTION WELL OIL AND WATER COMBUSTION GASES

① COLD COMBUSTION GASES
② OIL BANK (NEAR INITIAL TEMPERATURE)
③ CONDENSING OR HOT WATER ZONE (50° – 200°F ABOVE INITIAL TEMPERATURE)
④ STEAM OR VAPORIZING ZONE (APPROXIMATELY 400°F)
⑤ COKING REGION
⑥ BURNING FRONT AND COMBUSTION ZONE (600° – 1200°F)
⑦ AIR AND VAPORIZED WATER ZONE
⑧ INJECTED AIR AND WATER ZONE (BURNED OUT)

FIGURE 6.5 In-situ combustion is a major thermal means of oil recovery. Heat is generated in the reservoir by injecting air and burning part of the oil. This partially vaporizes the remaining oil, which is then driven forward by a combination of steam, hot water, and gas. Any oil left behind becomes fuel for the in-situ process. Water is also injected into the well; it improves the efficiency of the process by transferring heat from the rock behind the combustion zone (7) to the rock immediately ahead of the combustion zone (4). Reprinted with permission from *Enhanced Oil Recovery*. Copyright 1984 by the National Petroleum Council.

are located in both the western and eastern states. Eastern oil shales are intimate mixtures of inorganic silts and insoluble organic material that have been consolidated into rock. In western oil shales, the matrix is a carbonate-based marlstone. The organic content of both shales is typically 5–30 percent, and most of it is a polymeric, insoluble petroleum precursor called kerogen. Shale oil can be recovered by heating the rock through the range from 250 to 500°C, where the kerogen is thermally decomposed to liquid and gaseous products, leaving 20–35 percent of the organic matter as coke (Figure 6.7). Because of the insolubility of kerogen and the difficulty of physically separating it from the

shale, this retorting method is the only recovery process that has been developed.

Retorting can be carried out above ground or in situ. The former process involves mining the shale and heating it in a vessel. Process development has been concentrated on three important engineering problems. The first is how to handle the large quantities of solids that must be processed. The second involves how to transfer heat to those solids. And the third concerns the effect of shale particle size on the efficiency of oil recovery. In-situ retorting is an alternative to mining the shale, but only if the shale bed can be made sufficiently porous to allow injection of air to burn part of the kerogen and the resulting coke and to permit outflow of the retorting products. In shallow beds, blasting can lift the overburden and fracture the shale to permit these necessary flows. In deeper deposits, partial mining followed by blasting shale into the resulting space is used to create a porous rubble bed underground. Engineering research has focused on ways of producing porosity, on gas flow and combustion in porous beds, and on recovery of products from the large quantities of off-gas. Much of this research and development must be done in the field on a large and costly scale.

It is possible that greater porosity in shale beds could be achieved by chemical comminution of the shale. For example, the treatment of western oil shales with acid solutions might result in comminution by inducing corrosive stress fracture of the carbonate rock. Chemical engineering research in this area, as well in the elucidation of oil-rock interactions, might provide insights for new strategies for oil shale production.

Conversion of Coal to Gaseous and Liquid Fuels

Coal is the giant of fossil fuel resources. World reserves are many times those of petroleum, and the United States is one of the major resource holders. Coal can be used directly in combustion or converted to gas or liquid. Only combustion consumes significant amounts of coal today.

Coal is currently economically useful only in plants that are equipped for large-scale handling of solids, and it is used only indirectly as a raw material for chemical synthesis. Accordingly, there has been considerable research on processes for converting coal into gaseous or liquid fuels and chemicals. Only gasification has advanced to commercial status.

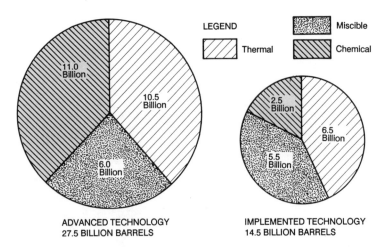

FIGURE 6.6 Prospects for enhanced oil recovery using implemented and advanced technologies are shown above. The ultimate amount of U.S. oil that can be recovered by "implemented technology," technology that presently exists in at least the proven field test stage, is estimated to be 14.5 billion barrels. Using "advanced technology," technology that might be conceivably developed before 2013, adds another 13 billion barrels of oil to the estimate, for a total of 27.5 billion barrels. A comparison of the distribution of ultimate recoveries by method is also shown. Most of the increase in the estimate from applying advanced technology comes from improvements in chemical flooding methods. The projections assume that crude oil has a nominal price of $30 per barrel and that the minimum rate of return on capital is 10 percent. Reprinted with permission from *Enhanced Oil Recovery*. Copyright 1984 by the National Petroleum Council.

Coal is gasified by heating it in the presence of steam to make synthesis gas (syngas), a mixture of carbon monoxide and hydrogen. A variety of processes for coal gasification have evolved, and several U.S. pilot facilities have been built on scales of 100 to 1,000 tons per day during the last 10 years. The Great Plains coal gasification plant in North Dakota is operating at 10,000 tons per day. Coal syngas has lower energy content than natural gas for fuel use, but is widely used for the synthesis of liquid fuels and other chemicals. For example, Tennessee Eastman is operating a commercial plant that converts 900 tons of coal per day into syngas that is in turn converted into acetic anhydride and other chemicals by a series of catalytic reactions. The Tennessee Eastman process is an excellent example of innovative chemical engineering in the design and construction of an efficient plant to synthesize organic chemicals from nonpetroleum raw materials. (See "Acetic Anhydride from Coal" on pp. 88–89.)

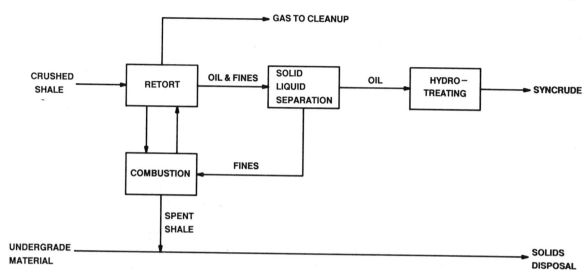

FIGURE 6.7 Steps in oil shale retorting are shown. Oil shale is crushed and then heated in a retort to drive off the oil that is trapped in the rock. Any oil left behind, as well as particulates returned to the process as the recovered oil is processed, is burned to provide heat for the retorting. The oil that is recovered from the shale is chemically treated to produce synthetic crude for further processing in conventional refineries. Courtesy, Amoco Oil Company.

Coal syngas can be converted into liquid hydrocarbon fuels by catalytic reactions. One process for this conversion, the Fischer-Tropsch process, was developed in Germany during World War II and is being operated on a large scale in South Africa. Today's pilot facilities and pioneering uses of syngas are establishing a technical and economic basis for the generation of commercial coal gasification projects that is expected to emerge in the 1990s.

Coal gasification research and development have concentrated on handling of solids, problems with ash, and dealing with the sulfur and nitrogen compounds present in coal. The newest pilot plants are investigating catalytic gasification. The integration of coal gasifiers with electric power generators in a combined-cycle mode (Figure 6.8) is an emerging field for design studies and economic evaluation. Much of the combined-cycle equipment is being perfected in natural-gas-burning plants that will come on stream in the next few years.

In-situ coal gasification has been demonstrated in small-scale field tests. Compared with aboveground gasification, in-situ gasification has the potential advantage of mitigating many of the problems associated with materials corrosion and mechanical solids handling; the main environmental problem is groundwater contamination. A further benefit of in-situ coal gasification is the potential for exploiting coal reserves that cannot be mined economically. However, it is no less complex than in-situ recovery of heavy tars or shale oil, and the engineering challenges are comparable.

The conversion of coal into liquid materials can be accomplished by pyrolysis or by direct liquefaction—heating coal in the presence of a hydrogen source. Neither of these routes is yet economically feasible.

In pyrolysis, coal is split into a hydrogen-rich liquid and a hydrogen-depleted solid char. The liquid contains significant amounts of nitrogen and sulfur compounds, as well as high-molecular-weight aromatic compounds such as asphaltenes. It is difficult to upgrade this liquid to a fuel suitable for transportation uses. Furthermore, while the liquid might be used in boilers, it would pose severe problems of nitrogen and sulfur oxide emissions from power plants. Research to date has uncovered few uses for the char, and it must be disposed of.

There are challenges and opportunities in developing a process for in-situ pyrolysis of coal in which the char is the principal fuel.

In direct liquefaction, coal is heated in the presence of hydrogen and a catalyst such as cobalt-molybdenum or nickel-molybdenum on alumina to give a greater yield of high-quality hydrocarbons than that produced by pyrolysis. This hydrogenation process has been demonstrated in several 50- to 250-ton-per-day plants.

Chemical engineering research on direct liquefaction has focused on improving hydrogenation efficiency, for example, by treating a coal slurry in a hydrogen-donor solvent in a high-pressure reactor. Basic knowledge of coal structure and reactivity as well as scientific understanding of hydrogen-transfer reactions has been crucial in improving the process. Equally important has been the realization that the desired reaction products can undergo secondary reactions that diminish yields and quality of final products. Accordingly, two-stage liquefaction with a catalyst in one or both stages is being tested on a small scale. Although catalyst performance is improving, there is still a need for catalysts that will perform even better in such severely fouling conditions.

New Raw Materials for Petroleum Refineries

As the domestic mix of fossil fuel resources changes over the coming years, new challenges will emerge for the design and renovation of our nation's installed base of refineries. While the practical aspects of this task must be left to the petroleum

Acetic Anhydride from Coal

The most successful example of generating chemicals directly from coal is the Tennessee Eastman integrated process for producing acetic anhydride. The commercial plant gasifies approximately 900 tons of coal per day and performs four chemical steps to yield annually 500 million pounds of acetic anhydride, 390 million pounds of methyl acetate, and 365 million pounds of methanol. In addition, 150 million pounds per year of acetic acid may be produced from acetic anhydride.

The process begins with a gasification process that converts coal into carbon monoxide and hydrogen. Part of this gas is sent to a water-gas shift reactor to increase its hydrogen content. The purified syngas is then cryogenically separated into a carbon monoxide feed for the acetic anhydride plant and a hydrogen-rich stream for the synthesis of methanol.

Eastman uses acetic anhydride primarily for esterification of cellulose, producing acetic acid as a by-product. This acetic acid is used to convert the methanol into methyl acetate in a reactor-distillation column in which acetic acid and methanol flow countercurrently.

The final step in the process involves reacting purified carbon monoxide from the gas separation plant with methyl acetate to form acetic anhydride, using a proprietary catalyst system and process. Part of the acetic anhydride is reacted with methanol to produce acetic acid and methyl acetate, and the latter is recirculated to the carbonylation step.

Eastman chemical engineers provided innovative solutions to key steps in the methyl acetate process. Computer simulations were used extensively to test ways to minimize the size of the reactors and recycle streams, to maximize yields and conversions, and to refine the methyl acetate in a minimum number of steps.

The novel approach finally taken was to conduct the reaction and purification steps in a reactor-distillation column in which methyl acetate could be made with no additional purification steps and with no unconverted reactant streams. Since the reaction is reversible and equilibrium-limited, high conversion of one reactant can be achieved only with a large excess of the other. However, if the reacting mixture is allowed to flash, the conversion is increased by removal of the methyl acetate from the liquid phase. With the reactants flowing countercurrently in a sequence of

and gas industries, there is a need for fundamental research to provide new design concepts and for trained engineering personnel to maintain international competitiveness in these inflashing reactor stages, the high concentrations of reactants at opposite ends of the column ensure high overall conversion.

The methyl acetate plant is highly heat integrated. Reboiler condensates are flashed to generate atmospheric steam for preheating the acetic acid feed, and the hot by-product wastewater is used to preheat the methanol feed. A vapor sidedraw provides most of the heat to run the impurity stripping column. Most importantly, the reactor column is inherently heat integrated. It can be thought of as four distillation columns stacked vertically, with the top and the bottom of each column acting as the reboiler and condenser for the columns above and below, respectively.

Because of the high level of heat integration and the combination of several unit operations in the methyl acetate reactor column, the design of a control system was a major challenge. The system had to take into account the broad range of time constants of the phenomena in the reactor column and the interaction of the column with impurity removal columns and heat integration devices. A control strategy was developed to solve these problems and to allow the operation of the reactor column at an exact stoichiometric balance of feed materials. Perhaps the most important testimony to the engineering development work is the fact that the commercial plant was a 500:1 scale-up of the largest pilot unit built.

The plant for the carbonylation of methyl acetate to acetic anhydride offered similar engineering challenges. Two pilot plants were operated to evaluate two competing schemes for reactor configuration. Early in the pilot plant stage, computer models for calculating detailed heat and material balances, operating costs, and capital requirements were programmed and continually updated from emerging pilot plant data. Equipment design was especially challenging because the reactor system scale-up factor was 10,000 times the pilot plant rate. The overall plant has several environmental advantages. The gasifier operates above 1,370°C, so no tar or other hydrocarbon waste is generated. The molten slag is quenched, crushed, and removed from the gasifier as inert granular pellets that qualify for nonhazardous landfill. Aqueous streams containing hydrocarbons from the chemical plants are recycled to the gasifier for slurry and quench water makeup. Tail-gas streams from the chemical plants are burned for heat recovery in existing boilers and furnaces. This plant is an excellent example of innovative chemical engineering in designing and building a chemically and thermally efficient plant to synthesize organic chemicals from nonpetroleum raw materials.

refineries have been designed, both heavy crudes and shale oil contain hydrocarbons of higher molecular weight and higher carbon-to-hydrogen ratio; more unwanted sulfur, nitrogen, and catalyst-poisoning metals like vanadium and nickel; and a bitumen-like residue that is difficult to refine. Such resources must be upgraded by chemically adding hydrogen or chemically removing carbon, by redistributing hydrogen among hydrocarbon fractions, and by removing compounds of heteroatoms and metals. Research has already led to improved thermal upgrading methods such as fluid-bed coking with coke gasification to remove carbon (Figure 6.9) and to catalytic hydrotreating schemes to add hydrogen and remove compounds of nitrogen, sulfur, vanadium, and nickel (Figure 6.10).

The co-processing of coal with heavy crude oil or its heavier fractions is being developed to lower capital requirements for coal liquefaction and to integrate processing of the products of coal conversion into existing petroleum refineries. This development appears to represent the main route by which coal-based liquid fuels will supplement and perhaps someday displace petroleum-based fuels.

At the low-molecular-weight end of the spectrum, a process newly commercialized by Mobil for converting methanol into gasoline has significantly expanded opportunities in C-1 chemistry— the upgrading of one-carbon molecules to multicarbon products. The process involves the use of ZSM-5, a shape-selective zeolite catalyst. (See "Zeolite and Shape-Selective Catalysts" in Chapter 9.)

Since methanol can be made from coal, nat-

dustries. The challenges arise from the properties of the new raw materials that are already finding their way into the process mix. In comparison with the crudes for which

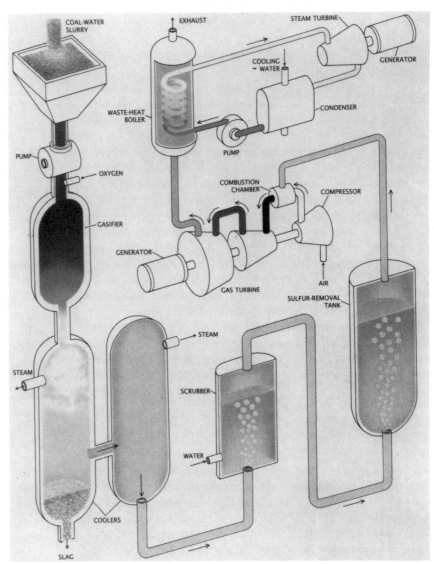

FIGURE 6.8 A process known as integrated gasification combined cycle (IGCC) is shown. It begins with the heating of a slurry of coal and water in an oxygen atmosphere. This produces a fuel gas composed mainly of carbon monoxide and hydrogen. After the gas has been cooled, cleansed of solid particles, and rid of sulfur it can be burned to drive gas turbines and then produce steam for a steam turbine. An IGCC plant emits fewer pollutants into the air than conventional coal-fired plants do. Reprinted with permission from *Scientific American*, 257 (3), September 1987, p. 106. Copyright 1987 by Scientific American, Inc.

ural gas, or even biomass, the methanol-to-gasoline (MTG) process establishes a new link between these resources and liquid fuels. The MTG process is now operating commercially in New Zealand (Figure 6.11), where a synfuels plant converts natural gas into high-octane gasoline at the rate of 14,500 barrels a day.[4] Further developments in the process promise to extend the product slate to include other fuels as well as lubricants and chemicals. Through the coal-

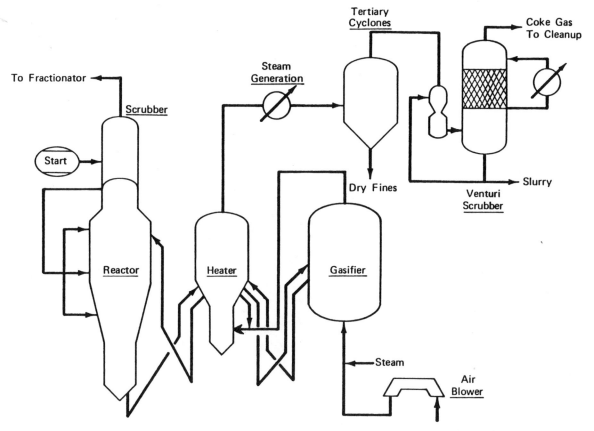

FIGURE 6.9 Flexicoking is a commercial process for refining petroleum that has been applied to heavy oil and tar sand fractions. The process employs circulating fluidized beds and operates at moderate temperatures and pressures. The reactor produces liquid fuels and excess coke. The latter is allowed to react with a gas-air mixture in the gasifier fluidized bed to provide a low-value heating gas that can be desulfurized and used as a plant fuel. Courtesy, Exxon Research and Engineering Company.

to-syngas route, this C-1 technology makes it possible for the United States to convert its largest natural resource, coal, into many of the liquid products that are now derived from petroleum.

There are a number of related challenges in research on new refining feedstocks. One is to find selective, economically viable catalytic methods to convert methane directly to liquids, thereby avoiding intermediate products such as methanol. A related challenge is to upgrade hydrocarbons in the range C-2 to C-4. These hydrocarbons are becoming more available as by-products from intensive refining processes used to make lead-free gasoline. Supplies of

these light hydrocarbons are likely to become increasingly abundant as their presence in gasoline is reduced by more stringent regulations to prevent hydrocarbon emissions to the atmosphere.

Municipal Solid Waste as an Energy Source

The use of municipal solid waste as a source of energy fills a special need by partially eliminating urban waste in an environmentally acceptable manner while at the same time producing usable energy.

Combustion of municipal solid waste, or of

FIGURE 6.10 Hydrotreating plant. Courtesy, Amoco Oil Company.

fuels derived from the waste, is a fledgling technology, especially in this country. There are currently 63 large plants for this type of combustion, and another 100 are starting up, under construction, or planned. Two different technologies are used. In one, which is employed by about 40 percent of these plants, the raw refuse is fed directly to the boiler; in the other, the refuse is first processed to reduce piece size and to remove noncombustibles. About one-third of the plants produce low-pressure steam for heating, another one-third produce high-pressure steam for electric power generation, and the balance also exhaust low-pressure steam for industrial or municipal heating.[5]

Process problems include slag formation, ash removal, and process control because of the heterogeneous solid waste feed. These problems have been managed to some degree by "over-designing" the plant, with the result that combustion of municipal solid waste is not economically competitive in areas where low-cost electricity or landfills for waste disposal are available. The future cost of electricity is difficult to predict. However, the steady decrease in the availability of landfills portends increasing use of this process to dispose of municipal wastes, particularly in large cities.

Nuclear Energy

The phrase "nuclear power" covers a number of technologies for producing electric power other than by burning a fossil fuel. Nuclear fission in pressurized water-moderated reactors—light water reactors— represents the current technology for nuclear power. Down the line are fast breeder reactors. On the distant horizon is nuclear fusion.

FIGURE 6.11 The world's first methane-to-gasoline plant, located in Motunui, New Zealand. In operation for about a year, this plant has already met its strategic objective of reducing New Zealand's dependence on foreign oil. Courtesy, Mobil Research and Development Company.

Nuclear Fission

Nuclear fission accounted for 13 percent of the electricity generated in the United States in 1985. Plants under construction in 1985 will probably raise the proportion to 20 percent by 1993. However, overexpansion of electrical generating capacity in this country, actual and imagined hazards of nuclear power plants, and negative perceptions of nuclear power by the public have combined to halt commitments to build new plants. New construction is not expected to resume before the 1990s.

Development efforts in the nuclear industry are focusing on the fuel cycle (Figure 6.12). The front end of the cycle includes mining, milling, and conversion of ore to uranium hexafluoride; enrichment of the uranium-235 isotope; conversion of the enriched product to uranium oxides; and fabrication into reactor fuel elements. Because there is at present a moratorium on reprocessing spent fuel, the back end of the cycle consists only of management and disposal of spent fuel.

Fast breeder reactors continue to be developed, although the level of support has fallen since the 1983 cancellation of the Clinch River Breeder Reactor project. Since that time, innovative fast reactor concepts like the Integral Fast Reactor (IFR) have made considerable headway. The IFR is a self-contained, sodium-cooled, metal-fueled reactor system. Its key features include a closely coupled fuel cycle for recovery, purification, and recycling of the uranium-plutonium core fuel alloy and extraction of a plutonium concentrate from the uranium blanket, where new plutonium is generated, for reenrichment of the core fuel. Any fast breeder fuel cycle must include fuel reprocessing because of the inescapably high concentration of fissionable materials in the used fuel. A novel aspect of the IFR reprocessing concept is the use of electrorefining rather than solvent extraction for recovery of fuel materials. Electrorefining technology will be carried out in a molten metal-halide salt electrolyte at about 500°C. To carry out electrorefining on a practical scale will require research and development in

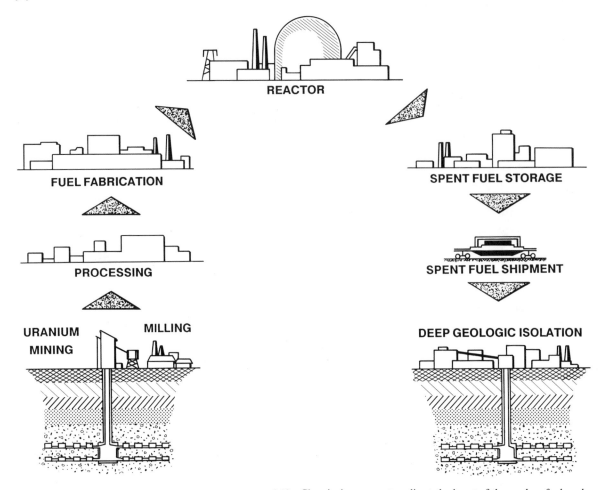

FIGURE 6.12 Chemical process steps lie at the heart of the nuclear fuel cycle used in the United States. Uranium from mining and milling operations is converted to UF$_6$, a volatile gas, which can then be separated so as to isolate the radioactive isotope of uranium. Further chemical steps are used to convert UF$_6$ into UO$_2$ for fuel assemblies, to isolate the long-lived isotopes generated in nuclear power plants, and to encapsulate these isotopes for eventual storage in nuclear waste respositories. Courtesy, Argonne National Laboratory.

such fields as pyrometallurgy, thermodynamics of metal-salt systems, ceramics, distillation of metals, waste processing, and remote process control technology.

Nuclear Fusion

Although generation of power from nuclear fusion has not yet been demonstrated, it is potentially a huge source of energy. Fusion power, the source of the sun's energy, results from the release of energy through the combi-

nation of light elements into heavier ones. The reaction currently under study is fusion of deuterium and tritium to produce helium and neutrons. Deuterium, an isotope of hydrogen present in low concentration in water, can be obtained by separating it from seawater. Tritium, another hydrogen isotope, is generated by reacting lithium with neutrons. In a fusion reactor, it would be generated in the lithium blanket of the reactor by neutrons derived from the reactor.

Major research programs in the United States

and the Soviet Union have concentrated on proof of principle and on containment of the extremely high-temperature nuclear reaction. Although the research is expensive, U.S. federal funding had been relatively ensured until recent budgetary problems. Chemical engineering research is needed on the preparation of solid breeder materials, blanket tritium recovery, blanket coolant technology, high-temperature heat transfer, low-activation materials, and tritium containment.

Electrochemical Energy Conversion and Storage

The last 25 years have witnessed a rapid rise in the numbers and capabilities of batteries, fuel cells, and electrolysis cells. Developments in batteries and fuel cells open the way to new and improved schemes for energy storage and power generation (see "Fuel Cells for Transportation" in Chapter 9). Devices invented for military and aerospace applications have moved quickly into civilian use—for example, in consumer electronic products, stationary energy storage, and electric vehicles. Significant advances in electric vehicle technology could significantly reduce demand for transportation fuels derived from petroleum.

The areas of important electrochemical engineering research can be grouped by the desired results. Initial costs can be lowered by devising better or less costly corrosion-resistant materials for inactive components of cells, superior catalysts for electrodes in fuel cells and electrolysis cells, and innovations that reduce the complexity of electrochemical transformation systems. The useful lifetimes of systems can be lengthened by reducing corrosion, improving electrodes, and understanding how porous electrodes perform and change in structure during cyclic operation. Finally, battery and cell performance can be improved by developing new materials that permit higher electrode reaction rates and lower internal electrochemical losses and by new designs based on modern electrochemical engineering.

Solar Power

The challenge of solar energy research is to discover or develop efficient processes for trans-

forming radiant solar energy into usable electrical energy or chemical fuels. Solar energy has inherent advantages over other energy sources. It is plentiful, ubiquitous, free, and continuously replenished, and it can be converted into electricity or fuels by processes that are environmentally benign. Yet, because it is a dilute and highly variable source of energy, conversion costs are high, and the construction of large (1,000-MW) commercial plants for electric power or fuels production is not likely in this century. Solar energy falls on the earth at about 1 kW/m^2 at noon on a sunny day. A solar cell area of about 40 km^2 (15.5 mi^2) would be required to construct a 1,000-MW power station with solar cells that are 12 percent efficient, assuming an average of 5 hours of sunlight per day throughout the year. The cost of siting and constructing a plant of this size, based on current technology, would be great. On the other hand, smaller photovoltaic arrays are being used as a peaking resource by some electric utility grids. For example, Southern California Edison maintains an array of about 0.1 km^2 to produce about 7.5 MW of power. It operates unattended and has low maintenance expenses.

Solar power research should be continued to make smaller scale applications of solar energy more cost-effective. Specific research areas for chemical engineers in photovoltaics include development of low-cost methods for producing the cell materials and for fabrication of the other components that are proportional to cell area. Solar-induced temperature gradients afford opportunity for ocean thermal energy conversion. Research is needed on convective and phase-change heat transfer as well as on biofouling control to minimize resistance to heat transfer on the seawater side. Development of low-cost materials and methods of construction for heat exchangers will be key to the success of this technology.

Research on the molecular basis of photoexcitation and electron transfer, including interactions of electron donor and acceptor molecules, could lead to new photochemicals. Development of model photosensitive compounds and methods of incorporating them into membranes containing donor, acceptor, or intermediate excitation transfer molecules, and eventual development of photochemical reactor systems, could involve the use of sunlight to replace less-specific energy sources.

Geothermal Energy

In some situations, it makes good economic sense to tap the earth's heat as an energy source. One approach to utilizing this resource is illustrated in the sidebar on this page.

Plant Biomass as a Fuel Source

During the 1970s, considerable attention was given to fuels from renewable resources. Use of wood-burning stoves increased markedly in some parts of the country. A subsidized industry to prepare ethanol as a gasoline supplement was started. But despite this, it is unlikely that plant biomass will ever satisfy more than a small percentage of U.S. energy demands. The reasons are primarily technological and logistical.

Much of the plant biomass produced on U.S. land and water areas is used for food or forest products or exists in designated-use areas such as parks and national forests. If 10 percent of the total biomass were available for conversion to energy, it would represent about 5.7×10^{18} J/yr (5.4×10^{15} BTU/yr). With the generous assumption of 50 percent conversion efficiency, this biomass would produce about 2.8×10^{18} J/yr (2.7×10^{15} BTU/yr) of energy as fuel, or only about 3 percent of current U.S. energy demand. Furthermore, the logistics of collecting today's available biomass would be forbidding. If a large area for producing biomass for energy were to be set up, it would almost certainly have to be on marginal land and would

Geothermal Energy

Geothermal energy is derived from the heat of the earth in reservoirs of rock-trapped fluids, often at depths of hundreds of meters. Its most common use is to power electrical generating plants. Over the past 20 years, geothermal energy technology has progressed to the point where it occupies a position of growing importance in the energy spectrum of many nations. Today, on a worldwide basis, geothermal energy powers close to 4,000 MW of installed electrical generating capacity.

Unocal Corporation is the world leader in the development of geothermal energy. Most of the technology involved in using geothermal energy has been adapted from the petroleum industry, and the use of several scientific disciplines has made it possible to meet the challenges of finding new geothermal sources and of moving quickly into commercial operation.

An example of the multidisciplinary approach used by Unocal to exploit geothermal energy is the development of the Salton Sea geothermal field in southeastern California. This resource is a brine at a temperature of over 260°C that contains more than 20 percent dissolved salts—nearly 10 times saltier than seawater. Wells drilled into the brine-bearing rock flow spontaneously, and the brine pressure is reduced, allowing part of the brine to flash to steam. The steam is routed through a turbine generator, and the spent brine and condensed steam are injected back into the geothermal reservoir.

Unique problems were encountered in drilling wells in the Salton Sea field. The hot salt solution in the reservoir adversely affected drilling fluids and equipment, requiring the use of specially formulated drilling muds and cements developed by chemists and of special steels recommended by corrosion and materials engineers.

Facilities for separating steam from brine and for power generation were designed by mechanical, chemical, and electrical engineers. Dissolved salts in the brine cause severe scaling and corrosion in wells and pipelines. Chemists and chemical engineers developed new production techniques to overcome these problems, as well as pollution control technology for the operation.

require greater than average use of fertilizer, irrigation, and mechanical work, all of which consume fuel energy themselves. The cost of net energy contributed by biomass will always be substantially higher than that calculable for the gross (i.e., apparent) fuel product. Thus, economic considerations would appear to rule out production of large amounts of energy from this source for the foreseeable future, and engineering research for such schemes does not merit high priority.

The development of bioreactor systems for the production of large-volume chemicals (see Chapter 3) could be the basis for reconsidering the production of biomass in limited quantities for fuel uses. This would require efficient microbial organisms to catalyze fermentation, digestion, and other bioconversion processes, as well as efficient separation methods to recover fuel products from process streams.

TECHNOLOGIES FOR EXPLOITING MINERAL AND METAL RESOURCES

The technologies involved in the minerals processing industry can be broken down into those where the desired metal component is in high concentration, such as scrap iron, iron ore, phosphate ore, and bauxite, and those where the concentration of the valuable constituent is low, such as gold and silver ore, lean copper ore, and certain types of scrap and wastes.

High-Concentration Raw Materials

The economics of extraction processes require the primary raw materials industries to locate near the richest ore deposits. Many such deposits are now outside the United States. The combination of high U.S. labor costs, foreign government subsidies, and very dilute domestic ore deposits (0.5 percent or less) has driven a substantial portion of the copper mining and refining industry to foreign countries. The development or acquisition of technology abroad for the smelting of nonferrous metal sulfides, electrolytic extraction of zinc, and production of steel has drawn portions of those industries overseas. Although U.S. industry has been slow to adopt the new technologies, there are indications that the pace of U.S. research and development in these areas may quicken. An example is the steel industry/federal government initiative in steel making. Projects now under way or planned include electromagnetic continuous casting, direct reduction of ore, and development of processes to remove copper and tin from scrap steel.

High concentrations of valuable elements exist in the earth's crust but cannot be economically recovered because they are buried too deep. The challenge to the chemical engineer is to develop methods for extracting these valuable materials in place without having to move and process enormous amounts of rock. The general concepts for in-situ recovery by solution mining or leaching are practiced for the recovery of uranium, soda ash, and potash. Despite these successes, most of the opportunities remain untapped because of technological barriers. Each mineral deposit has its own characteristics, and the processing environment deep beneath the earth's surface has been constructed by nature; our ability to modify it is limited.

Many research needs in this area parallel those of in-situ processing of oil shale and recovery of heavy crude oils. Additional research needs cover a broad spectrum of mining, metallurgical, environmental, and chemical engineering: solids handling and comminution, separations and concentration processes for ore beneficiation, electrolytic processing, solvent extraction, and treatment and disposal of waste products. It will take the best-trained chemical engineers using the most sophisticated tools of chemistry, physics, and computer technology to unlock economically the vast reserves of metals and minerals that cannot be recovered at present.

Low-Concentration Raw Materials

The ascendant method for economically processing deposits low in the desired component is solvent extraction. This is most commonly done in processing plants above ground, and all the spent ore must be restored to its original location or otherwise disposed of in a way that meets environmental constraints. The restoration cost can be borne by high-value products like gold, silver, and uranium. In the production of moderate-value products such as copper from lean ores, magnesium from dolomite, and aluminum from raw materials other than bauxite, the only economically viable way to process deposits may be to extract more than one product. This requires designing and building more complex chemical plants, with all the attendant challenges to chemical engineering research.

Waste Streams as Sources of Minerals and Metals

Substantial quantities of aluminum, copper, and steel are reused as scrap. The challenge is to purify the scrap metal sufficiently to process it for reuse. There is opportunity for new processes that can remove unwanted elements—either alloyed or piece contaminants—more effectively and at lower cost than current processes.

Many of the waste streams from U.S. process industries are water containing small quantities of metal ions that the law requires be removed before the wastewater is disposed of. There is an economic incentive to recoup at least some of the cost of wastewater treatment by recovering and selling the metal content instead of merely disposing of the metals as sludge. Because the waste streams are dilute in desired materials, research is needed to devise efficient extraction and separation processes.

Likewise, fly ash from power plant combustors often contains small amounts of metals or their oxides, which require costly disposal in the ever-shrinking number of approved hazardous waste landfills. Thus, there are economic incentives to recover the metal values as well as to reduce the costs of ultimate disposal. Here, too, the metal content is low, and research is needed to develop economical separation processes. In principle, advances in this area could be translated into recovery of metal values from mine tailings.

INTELLECTUAL FRONTIERS

The basic technologies used in the energy and natural resource processing industries have many elements in common, and the chemical engineering profession has a long history of finding and adapting basic technologies to the needs of diverse industries. No profession is better suited by tradition and training to attack the many difficult technical problems of these industries. And these problems must be attacked and solved if our country is to maintain its high standard of living and its position in the worldwide economy.

The demands for energy and materials continue to increase, and the accessibility of natural resources to meet them continues to fall as the most easily recovered fossil fuel and mineral deposits are depleted. The gap between rising demand and falling availability must be bridged by technology that improves the efficiency of extraction, conversion, and use of energy and materials. The development of such technology takes long lead times, and there is a paramount long-term need to maintain momentum in research on the frontiers of chemical and process engineering. The problems enumerated here offer challenges equal to any that chemical engineers have faced in the past.

In-Situ Processing

Available resources of fuels and materials in the accessible parts of the earth's crust are becoming increasingly scarce. The alternative to moving greater and greater amounts of crust, whether it is mixed with the valued substance or simply overlies it, is in-situ processing. Although this technology is well established in petroleum recovery, the long-term incentive to increase its efficiency is great. The incentives for other in-situ technologies vary but are bound to intensify in the future. The development of in-situ processes involves long lead times in research and development. Field tests are large-scale, prolonged projects that may last many months. The potential environmental problems are considerable. By the time the need for an in-situ process becomes acute, it is too late to commence research. The prize goes to those who are prepared.

Problems with in-situ processing share certain elements. Fluid phases move through a vast, complex network of passages in a porous medium. The process is inherently nonsteady state. The physical transformations or chemical reactions proceed in zones or fronts that migrate through the porous structure. The fluids interact physically with the solid walls that define the passages. The passages are irregular, and their dimensions and structure change with distance. This structural inhomogeneity imposes uncertainties that make processing in situ riskier than processing in designed and constructed plants. Further, the potential adverse environmental impacts of in-situ processing have proved to be important barriers to the widespread commercialization of in-situ processes for oil shale and coal. Sustained research in the following areas is needed to reduce both environmental and process risks:

- porous structures, both at the microscopic scale and larger;
- methods for creating or enhancing permeability in nonporous formations of oil shale, coal, and ore bodies;
- combustion processes under reservoir conditions;
- mechanisms of oil displacement;
- the distribution and flow of viscous fluids in porous media and the motion of complex fronts;
- surface and colloidal phenomena involved in fluid-rock interactions, such as wetting and spreading, and adhesion and release;
- phase equilibria, phase thermodynamics, and chemical reactions between injected fluids and solids in the reservoir;
- phase behavior, colloidal aspects, adsorption, and rheology of surfactant formulations;
- rheology and degradation of hydrophilic polymers and their interactions with rock;
- the chemistry involved in winning a desired component from a given type of deposit;
- separation of fines from produced materials;
- treatment and disposal of tailings;
- mathematical models of such phenomena; and
- process synthesis, design, management, and optimization with severely limited information.

Geochemistry, geophysics, geology, environmental science, and chemical engineering must be more closely linked if advances are to be made in these areas.

Processing Solids

Solids handling is ubiquitous in the processing of energy and natural resources. To liberate the desired components, crystalline solids (e.g.,

rocks) must be broken into grains; these may have to be comminuted to yet finer particles. Current crushing and grinding processes are highly energy inefficient; typically 5 percent or less of the total energy expended is used to accomplish solids fracture. These processes also produce a broad distribution of particle sizes, including fines that are difficult to process further. Solids comminution could be greatly improved by a process that fractured crystalline solids selectively along grain boundaries.

Fundamental understanding of crushing, grinding, and milling is deplorably limited. For example, there is no rational basis for the design of a ball mill, a commonly used industrial device. Mineral processing and chemical engineering researchers need a deeper understanding of solid-state science and fracture mechanics, just as they have mastered and are contributing to colloid and interface science. There are great challenges in devising chemical comminution aids as well as processes for handling solids that become plastic, sticky, or reactive at temperatures reached in comminution.

Just as important as finding better ways to prepare granular solids and powders is finding ways to move them, to contact them with fluids, to allow them to react in chemical processes, and to separate the residues. A major study in 1981[6] showed that cost overruns on large projects involving solids processing depended directly on the throughput rate of solids. Much of the current equipment design for mineral processing dates from earlier times when ores were richer and costs of processing not as high.

The handling of coal, oil shale, and ores would be improved by research on the mechanics of pneumatic and slurry transport of particulate solids, particularly on the mechanisms of failure through plugging, attrition, and erosion. Improved processes for coal liquefaction and gasification could come from research on particulate transport in fluidized beds, including high-pressure gas-fluidized beds of large particles, ebulated beds, and liquid slurry reactors. We must also understand chemical reaction processes in systems of moving particles, especially at high temperatures and pressures. There are the related critical issues of particles being consumed or created by chemical reaction, particle agglomeration and sintering, and transport and separation of hot sticky particles. For example, Plate 5 shows the stages of retorting of an oil shale particle as the temperature is increased by an external hot gas. The kerogen is reacted, cracked, volatilized, and coked in an idealized series of concentric volumes. Liberated products must flow through the coked zone to exit the particle. If the temperature falls too rapidly, particles can become wet and sticky. These problems are related to the more general problem of chemical reactions involving liquids or gases inside porous solid particles.

Equipment design and scale-up present particularly great challenges whenever solids are to be processed on a large scale. Consequently, advances in the basic understanding of solids processing will be for naught if they are not translated into practical, reliable designs. This will require close cooperation among the fields of mechanical, mineral, and chemical engineering and between disciplines in the earth and physical sciences.

Separation Processes

Separations play a vital role in the processing of energy and natural resources.[7] Improved separations can lead to improved efficiency of existing processes or to economical means for exploiting alternative resources. For example, the petroleum refining industry is based on separations of natural and synthetic hydrocarbons. Improved separations could lead to better concentrations of aromatic hydrocarbons in gasoline to enhance the octane rating and paraffinic hydrocarbons in jet fuel to improve burning characteristics. The winning of critical metals such as copper, uranium, and vanadium from low-grade domestic ores requires chemical extraction followed by recovery from the dilute extractant solution. More selective extractants are needed, as are better separations to remove fly ash, sulfur oxides, and nitrogen oxides from power plant and other gaseous emissions to protect air and water quality.

Every separation process divides one or more feeds into at least two products of different composition. Separation processes that operate on heterogeneous feeds usually involve screen-

ing or settling. Those that involve physically homogeneous mixtures must use more subtle means to create products of different composition. These latter processes are pervasive in industry; they consume large amounts of energy and require sophisticated research and design.

Separation processes are based on some difference in the properties of the substances to be separated and may operate kinetically, as in settling and centrifugation, or by establishing an equilibrium, as in absorption and extraction. Typical separation processes are shown in Table 6.1. Better separations follow from higher selectivity or higher rates of transport or transformation. The economics of separation hinges on the required purity of the separated substance or on the extent to which an unwanted impurity must be removed (Figure 6.13).

Most methods of separating molecules in solution use direct contact of immiscible fluids or a solid and a fluid. These methods are helped by dispersion of one phase in the other, fluid phase, but they are hindered by the necessity for separating the dispersed phase. Fixed-bed adsorption processes overcome the hindrance by immobilizing the solid adsorbent, but at the cost of cyclic batch operation. Membrane processes trade direct contact for permanent separation of the two phases and offer possibilities for high selectivity.

There is already intensive research on membrane separations for energy and natural resource processing. Applications have so far centered on organic polymeric membranes for mild service conditions, but research could lead to both organic and inorganic membranes that can operate under harsher conditions. Zeolites and other shape-selective porous solids like pillared clays appear to offer a fertile field of research for separation applications. Chemically selective separation agents that distinguish between absorptive, chelating, or other molecular properties are also attracting study.

Research should continue on traditional separation methods. For example, there is a continuing need for more selective extraction agents for liquid-liquid and ion-exchange extractions. High-temperature processes that use liquid metals or molten salts as extraction agents should have potential in nuclear fuel reprocessing and

TABLE 6.1 Methods for Separating Mixtures

Property Difference	Examples of Processes
Particle size	Screening, mechanical jigging
Magnetism	Magnetic separation
Density	Centrifugation, settling, jigging
Solubility	Extraction
Surface affinity	Adsorption
Solid/liquid phase	Filtration
Molecular character	Dialysis, membrane gas separation
Molecular size	Molecular sieve separation
Dielectric constant	Electrophoresis
Solidification temperature	Zone refining
Rate of phase change	Crystallization, distillation
Ionic character	Ion exchange

metals recovery; basic thermodynamic data on such high-temperature systems are lacking.

Many of the ores of base metals are sulfide deposits. They must be milled to exceedingly fine size in order to free the wanted grains from the rest of the mineral. The desired grains are semiconducting colloidal particles, and the mechanisms of leaching and flotation—the pre- ferred methods of concentrating them—depend on both their electrochemical and colloidal prop- erties. The separation processes leave a large quantity of unwanted fines that must be rejected as slimes. Better understanding of these pro- cesses should permit separation of complex sulfides and discovery of paths to recovering individual metals from dilute, impure solutions.

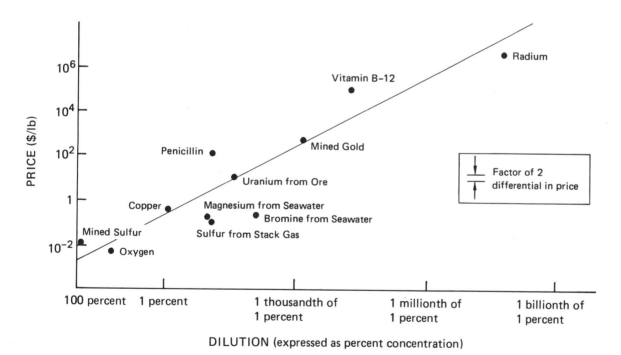

FIGURE 6.13 The importance of separation processes in determining the eventual cost of materials and products is illustrated in this figure. Product prices correlate with the degree of dilution of the raw material in the matrix from which it must be isolated. A factor of two in product price is shown in the figure. Courtesy, Norman N. Li, Allied-Signal Corporation.

A hypothetical separation of a homogeneous mixture, carried out in a thermodynamically reversible manner, would require the theoretical minimum expenditure of energy. In practice, however, separations of such mixtures need 50 to 100 times this minimum. Thus, there is significant opportunity for improvement of separations by creating ways to reduce energy consumption without a commensurate increase in capital and operating costs.

Researchers in separation science and technology draw on and contribute to a variety of related fields, including

- phase-equilibrium thermodynamics;
- mass transfer and transport phenomena;
- interfacial phenomena, including surface and colloid chemistry;
- mechanisms of chemical reactions, especially complexation reactions;
- analytical chemistry; and
- computer-assisted process and control engineering.

Future progress in separation science and technology will require continued cooperative research between scientists and engineers in these fields.

Materials

Research on materials can lead to more economical processing under extreme conditions and to reduced capital and operating costs. There are strong incentives to find construction materials for process units that are derived from domestic resources, that are less contaminating of process and environment, and that have the following properties:

- greater strength and more resistance to abrasion and corrosion;
- longer life and less subject to degradation by cycling conditions;
- serviceability under more severe conditions of temperature, pressure, or neutron flux; and
- greater resistance to hydrogen embrittlement.

There are comparable incentives to develop new process-related materials that are more selective as catalysts, extractants, or separation membranes and more effective in controlling flow in porous media. In addition, the development of materials that are less energy intensive in terms of production and use is a goal equivalent to other means of energy conservation.

The relatively mature technology of upgrading heavy oils by reaction with hydrogen is illustrative. Reactors are required to withstand hydrogen embrittlement at high pressures and temperatures. Present practice is to use foot-thick reactors lined with alloy steels. The largest of these can no longer be constructed in the United States because the cessation of nuclear power construction has led to the closing of facilities capable of such fabrication. Cheaper reactor materials would improve the economics of the process; better materials could lead to operability under more severe conditions that would provide higher conversions.

Materials problems abound in the energy storage field. For example, cheap materials with large effective thermal capacity are needed to store thermal energy in solar heating systems. Some systems use chemically reactive materials and store energy as enthalpy of reaction. There is an opportunity to develop photosensitive catalysts to improve the coupling between the solar energy input and the energy converter. High-temperature thermal energy storage systems confront corrosion problems aggravated by thermal cycling and temperature-sensitive solubilities that conspire to shorten system life, signifying the need for better materials. Likewise, battery storage of electrical energy is limited by the cost of materials and by corrosion and microstructural changes aggravated by the inherently cyclic operation.

Materials science is an intrinsically interdisciplinary field. Materials scientists include physicists, chemists, metallurgists, mechanical engineers, and chemical engineers. It is the latter who have the best opportunity to establish specifications for needed materials and to join in research on ways to meet those specifications.

Advanced Methods for Design and Scale-up

Many of the shortcomings of energy and natural resource processing arise from lack of sufficiently powerful design and scale-up procedures for the practicing chemical engineer (see Chapter 8). A goal of research is to design large units from first principles and small-scale experiments. This has been done in the past; scale-up factors of 50,000 are common in petroleum refining technology. However, in much of energy and natural resource production, there is such complexity and lack of basic data, especially for large-scale solids processing, that empiricism will continue to prevail until pilot plants and demonstration projects are successfully modeled. Scale-up factors in solids processing typically range from two to five.

For example, research on moving hot solids will require individual pieces of equipment, then whole systems, from which reliable data for scale-up can be obtained. Costs of such research would be out of reach for all but the largest industrial and government laboratories. The application of research results to improved commercial oil shale retorting would require large pilot plants costing tens of millions of dollars, followed by demonstration plants or single commercial modules costing hundreds of millions. Experimentation on an equivalent scale can be imagined for a new steel-making technology, for in-situ leaching of uranium, or for solution mining of hydrothermal mineral deposits such as soda ash. Such research will require interdisciplinary teams and sustained activity over periods of years.

The problem of ever-increasing construction costs dates from the mega-project concept of World War II and the race toward an overnight synthetic fuels industry. Nevertheless, large construction projects will be needed to bring coal gasification, coal liquefaction, and oil shale processing to fruition. Construction costs are not small in the minerals processing industries, and the more dilute the ore, the larger must be the economically viable plant. The incentives are great to develop lower cost designs and construction methods. Noteworthy ideas include modular construction from preassembled units and the organization of construction into multiple small projects. Substitute materials that can be produced with lower energy or raw material cost need to be developed. Chemical engineers must be cognizant of the construction cost implications when they select construction materials and prepare flowsheets.

Other Important Research

Many additional intellectual challenges for chemical engineers are relevant to energy and natural resource processing. These include reservoir modeling (Chapter 8), combustion (Chapter 7), catalysis (Chapter 9), and electrochemical engineering (Chapter 9). A recent report entitled *Future Directions in Advanced Exploratory Research Related to Oil, Gas, Shale, and Tar Sand Resources*[8] discusses some of the chemical engineering research challenges described in this chapter in a broad, multidisciplinary context that includes the earth sciences.

Finally, it should not be forgotten that chemistry plays an important role as a fundamental science for these industries. Major contributions to be expected from chemistry in energy and natural resources are discussed in the 1985 report *Opportunities in Chemistry*.[9]

IMPLICATIONS OF RESEARCH FRONTIERS

Each of the generic research areas discussed in this chapter has a strong multidisciplinary character. While the underlying fundamentals of some are amenable to investigation by individual chemical engineers, in many cases collaboration will be required between chemical engineers and other scientists and engineers skilled in geology, geophysics, hydrology, mechanical engineering, physics, mineralogy, materials science, metallurgy, surface and colloid science, and all branches of chemistry. It will be necessary to generate creative interactions that overcome traditional academic compartmentalization of outlook, experience, and education. Academic departments of chemical engineering should take the lead in establishing interdisciplinary teams to carry out fundamental research in these high-priority research areas. They should seek ways to involve government and industrial scientists in interdisciplinary activities. There must be freer flow of information between industry, university, and government; professional disciplines; and academic departments.

The educational background of chemical engineers makes them particularly well suited to solve problems in the areas discussed herein. Chemical engineers are used to working with concepts from all the related fields, and their training has evolved to cover most of the skills needed to solve technical problems. Interdisciplinary research in the relevant areas can only strengthen the chemical engineering cadre in the energy and natural resource processing industry.

The funding needs for the research described in this chapter will be large and long term; they can be met only by some combination of government and industry. Government support is appropriate because efficient processing of energy and natural resources is key to continued national growth and prosperity. Appropriate initiatives for the Department of Energy, the U.S. Bureau of Mines, and the National Science Foundation—all in cooperation with industry— are laid out in Chapter 10. Industry support is mandatory because commercialization is a goal and because companies will be the eventual profit-driven proprietors of the technology developed. There is advantage and precedent for companies to band together in consortia or through institutes to provide continued funding, particularly of basic research. In addition, such long-term commitment will allow academic researchers the freedom to set up ongoing programs to feed basic data and concepts into the centers and consortia without fear of sudden shifts in funding priorities.

NOTES

1. U.S. Department of Commerce, Bureau of the Census. *Statistical Abstract of the United States: 1987*, 107th ed. Washington, D.C.: U.S. Government Printing Office, 1986, Table 1310.

2. U.S. Department of Commerce, Bureau of the Census. *Statistical Abstract of the United States: 1987*, 107th ed. Washington, D.C.: U.S. Government Printing Office, 1986, Tables 1233, 1215.

3. National Petroleum Council. *Enhanced Oil Recovery*. Washington, D.C.: National Petroleum Council, 1984.

4. J. Haggin. "Methane-to-Gasoline Plant Adds to New Zealand Liquid Fuel Resources." *Chem. Eng. News*, 65 (25), 22 June 1987, 22.

5. E. Berenyi. "Overview of the Waste-to-Energy Industry." *Chem. Eng. Prog.*, 82 (11), November 1986, 13.

6. E. W. Merrow et al. *Understanding Cost Growth and Performance Shortfalls in Pioneer Processing Plants* (Report R-2569–DOE). Santa Monica, Calif.: Rand Corporation, September 1981.

7. This section draws, in part, on a recent NRC report, *Separation and Purification: Critical Needs and Opportunities*. Washington, D.C.: National Academy Press, 1987.

8. National Research Council, Board on Chemical Sciences and Technology. *Future Directions in Advanced Exploratory Research Related to Oil, Gas, Shale, and Tar Sand Resources*. Washington, D.C.: National Academy Press, 1987.

9. National Research Council, Board on Chemical Sciences and Technology. *Opportunities in Chemistry*. Washington, D.C.: National Academy Press, 1985.

SEVEN

Environmental Protection, Process Safety, and Hazardous Waste Management

Chemical engineers face important research challenges associated with the imperative to protect and improve the environment. These challenges include designing inherently safer and less polluting plants and processes, improving air quality through research on combustion, managing hazardous wastes responsibly, developing multimedia approaches to the study and control of pollutants in the environment, and assessing and managing chemical risks to human health or to the environment. The future of the profession and the industries that it serves will depend on the vigor with which chemical engineers approach these challenges. These challenges have important implications for the way that chemical engineers are educated and for the means by which government and industry provide research support to the field.

A IR, SOIL, AND WATER are vital to life on this planet. We must protect these resources and use them wisely—our survival as a species depends on them. Despite recent impressive strides in improving the environment, evidence is overwhelming that more effective action must be taken to address such critical issues as acid rain, hazardous waste disposal, hazardous waste landfills, and groundwater contamination. It is also vital that we assess realistically the potential health and environmental impacts of emerging chemical products and technologies. The problems are clearly complex and demand a broad array of new research initiatives.

Technological activities—of which chemical manufacture and processing are key parts—begin with the extraction of raw materials from the environment. They then proceed through numerous steps—processing, storage, handling,

transportation, and use—and finally end with the ultimate return of processed materials or their residues to the environment (Figure 7.1). The resulting redistribution of chemicals within the environment may have adverse impacts. Harmful compounds of certain elements (e.g., nitrogen, sulfur, halogens) may be widely mobilized. Other elements may be converted from innocuous forms (e.g., mercuric sulfide in cinnabar) to highly toxic forms (e.g., methyl mercury). Chemical engineers are involved in all aspects of chemical manufacture and processing; therefore, it should be their responsibility to safely manage chemicals in the environment. Environmental and safety concerns will be crucial challenges to the chemical engineers of the future. No other group is better trained or more centrally positioned in the industrial world to be the cradle-to-grave guardian of chemicals.

Our approach to environmental and safety

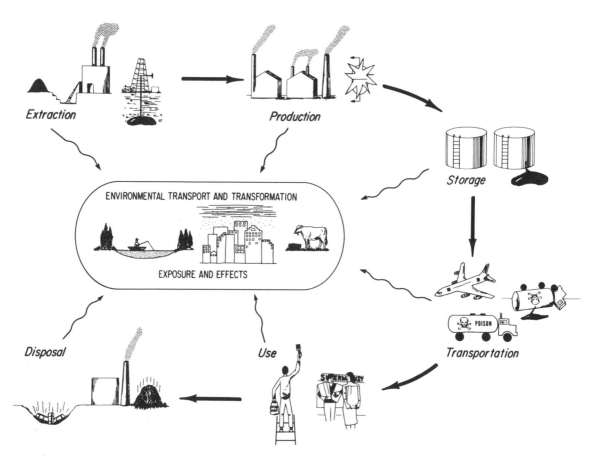

FIGURE 7.1 Life cycle of chemicals in the environment.

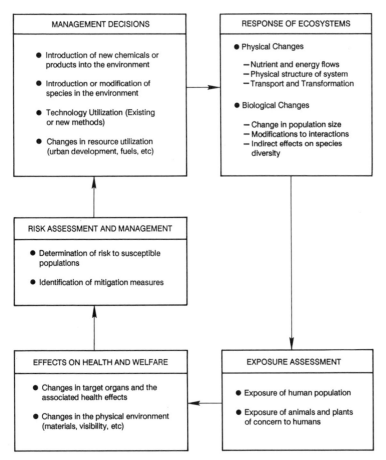

FIGURE 7.2   The effects of human activities on the environment.

in turn affect human beings and other resources. A risk assessment of the potential health and welfare effects of such changes can indicate whether and to what extent mitigation measures should be taken, or original decisions rethought. Each of the boxes in the diagram, although greatly simplified, represents a cluster of questions that require answers from environmental research.

IMPACT ON SOCIETY OF CHEMICALS IN THE ENVIRONMENT

A number of environmental issues have received widespread publicity (Table 7.1), from major accidents at plants (e.g., Seveso and Bhopal) to the global and regional impacts associated with energy utilization (e.g., carbon dioxide, acid rain, and photochemical oxidants), the improper disposal of chemical waste (e.g., Love Canal and Times Beach), and chemicals that have dispersed and bioaccumulated affecting wildlife (e.g., PCBs and DDT) and human health (e.g., cadmium, mercury, and asbestos).

As a consequence, much of the public has come to believe that most chemicals are hazardous. A recent poll by the Roper Organization revealed that two out of three American citizens expect a major chemical disaster, resulting in thousands of deaths, within the next 50 years.[1] The poll also found that a high proportion of the public lacked confidence that industry would deal openly with them. A public attitude toward exposure to chemicals is developing that can be summed up by the words, "no risk." But, as a judge recently stated, "In the crowded conditions of modern life, even the most careful person cannot avoid creating some risks and accepting others. What one must not do, and what I think a careful person tries not to do, is to create a risk which is substantial."[2]

problems must change from reactive to proactive; in other words, instead of responding to crisis and public pressures, we must anticipate and prevent problems. This will require an understanding of the detailed chemistry and physics of processes at the molecular level. Such basic understanding is crucial if we are to design plants that are safer and that cause less pollution, develop better ways to manage and detoxify hazardous waste, and predict the fate of chemicals in the environment as well as the effects of chemicals on humans and ecosystems.

Figure 7.2 diagrams how human activities can affect environmental quality and human health. Introduction of new chemical products, adoption of different technologies, or changes in resource utilization can lead to emissions that affect the physical, chemical, or biological responses of the receiving ecosystems. This can

TABLE 7.1   Some Well-Publicized Environmental Issues

Global	Chlorofluorocarbons and their effect on ozone in the upper atmosphere
	Carbon dioxide and other "greenhouse" gases (e.g., methane) and their effect on global temperature
	DDT
	Polychlorinated biphenyls (PCBs)
Regional	Acid rain (NO_x, SO_x)
	Agricultural wastewaters (Kesterson Wildlife Refuge, Chesapeake Bay)
Urban	Airborne lead from automobile exhausts
	Carbon monoxide from automobile exhausts
	Photochemical oxidants, particularly ozone (Los Angeles)
	Sulfur oxides (SO_x) and particulate matter (London, Donora)
Site-specific	Dioxins (Seveso, Love Canal, and Times Beach)
	Methyl isocyanate (Bhopal)
	Methyl mercury (Minamata)
	Cadmium (Itai-Itai)
	Indoor air (Formaldehyde, asbestos, NO_2, CO)

What is a substantial risk? How safe is safe enough? These are questions that trouble the public, industry, and regulators alike. Scientific understanding and the available data are inadequate to evaluate the true risk to individual safety, or the true risk of damage to human health or the environment, from exposure to most chemicals or to chemical plant or disposal operations. Yet, legislators and regulators cannot wait for all the data to come in before they start to provide the public with the protection it is demanding. Laws have been passed and regulations have been developed that require government approval for production of new chemicals, design and operation of chemical plants, workplace exposures, certain product uses, quantities and concentrations of chemicals in effluent streams, and disposal of waste and by-product streams. But because of the uncertainties, some regulations have been written to protect against the effects of extremely unlikely worst-case scenarios, resulting in a misallocation of resources, reduced technical innovation, and excessive costs. At the same time, other, less visible hazards that might be the focus of appropriate regulation have been overlooked.

This situation must be corrected. Society needs a clean and safe environment. It also needs to capitalize fully on new developments in chemistry, biotechnology, and materials science, if the United States is to retain techno-logical leadership and international competitiveness in major segments of industry.

Some of the economic and social costs related to major environmental issues are discussed in the following sections.

Chemical Industry Safety

The U.S. chemical and petrochemical industry safety record is generally good. The National Safety Council's 1985 data show that the chemical industry worker is only one-fourth as likely to have a fatal on-the-job accident as the average U.S. employee.[3] The Bureau of Labor Statistics indicates that the 1984 chemical industry rate for lost work days resulting from occupational injury was 2.4 per 200,000 man-hours, compared to 4.7 for manufacturing as a whole and 3.7 for all private sector employment.[4] Over the past decade, annual fatalities from hazardous chemical accidents have numbered around 40, while highway fatalities have numbered around 40,000.[3]

Nevertheless, accidents and unintended chemical releases pose serious financial risks to the chemical and petrochemical industry. In 1984 there were five major accidents in the hydrocarbon-chemical industries, totaling an estimated loss of $268 million.[5] Hundreds of lesser accidents occur yearly. The total annual cost to the industry of accidents and unintended chemical releases is difficult to quantify. It includes significant costs owing to interruption

FIGURE 7.3 In 1984, 14 years after the passage of the Clean Air Act, significant areas of the United States were still in violation of National Ambient Air Quality Standards (NAAQS) for ozone. Courtesy, Environmental Protection Agency.

of business as well as major liability and litigation costs associated with injuries, deaths, property damage, and insurance premiums. It also includes losses of product and feedstock that are direct profit losses for the manufacturer. One estimate is that U.S. industry spent $7.7 billion in 1985 for protecting worker safety and health;[6] the total annual cost of accidents and unintended chemical releases by the U.S. chemical and petrochemical industries is surely many billions of dollars.

Costs associated with increased government regulation are also difficult to quantify. Public concern in response to chemical release accidents affects regulators and community policy groups. It is evident that the U.S. chemical industry is already spending large amounts of money to avoid accidents and to deal with their consequences when they occur; these costs are borne in part by the consumers. Continued expenditures are likely as industry strives to achieve an ''acceptable'' level of public safety throughout all chemical industry operations.

Combustion of Fuels for Power Generation and Transportation

The burning of fuel for power generation and transportation presents some of the most long-

standing and important problems in environmental protection. Fossil fuels are used in such magnitude that emissions from combustion sources have a major impact on urban, regional, and global air quality. Combustion-generated pollutants are derived from contaminants in the fuel such as sulfur, nitrogen, and inorganic compounds, from incomplete combustion of the fuel, and from the high-temperature reaction of nitrogen with oxygen in the air heated by combustion processes. These pollutants are emitted both in gaseous form (e.g., the oxides of nitrogen—the sum of NO and NO_2, denoted as NO_x—sulfur dioxide, and unburned hydrocarbons) and in particulate form (e.g., fly ash and soot). The 1970 Clean Air Act and its amendments are directed at reducing combustion-generated emissions through the establishment of National Ambient Air Quality Standards (NAAQS) for oxides of sulfur and nitrogen, carbon monoxide, particulate matter, lead, and ozone. While substantial reductions in urban levels of carbon monoxide, sulfur oxides, and particulate matter have been achieved over the past 15 years, ozone levels, controlled by an intricate chemistry involving organic gases and oxides of nitrogen, have proved to be more resistant to control (Figure 7.3). Large-scale air quality problems arising from combustion-generated emissions, such as acid rain (Figure 7.4), regional hazes, and volatile toxic compounds, have also assumed prominence and will likely be the targets of future legislation.

Adding fluidized gas desulfurization to coal-fired generating plants is estimated to add 15–25 percent to their total capital costs, or up to $125 million on a typical 500-megawatt unit.[7] Since the cost of reducing emissions by modifying the combustion process is usually an order of magnitude lower than that of cleaning the fuel before burning or removing the pollutants from the exhaust gases, there are significant challenges to develop clean, fuel-efficient com-

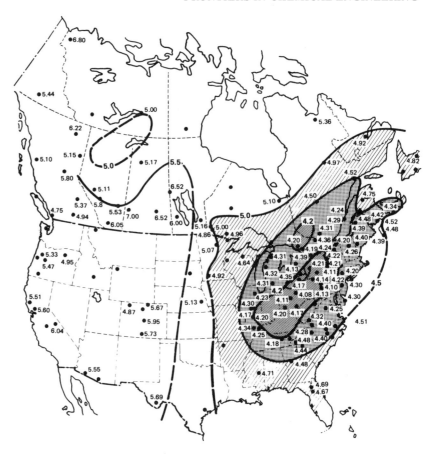

FIGURE 7.4 Annual mean value of pH (acidity) in precipitation, weighted by the amount of precipitation in the United States and Canada for 1980. The low pH values seen in the eastern United States and Canada reflects the high acidity of the rainfall and other precipitation in this region. This geographic pattern of acid rain correlates with the emission and transport of sulfur dioxide from coal-burning plants. From U.S./Canada Work Group #2, *Atmospheric Science and Analysis*, U.S. Environmental Protection Agency, Washington, D.C., 1982.

bustion processes as well as to design more economical processes for fuel cleaning prior to combustion and for destroying or removing the residuals from postcombustion gases.

Hazardous Waste Management

The disposal of hazardous waste may well have become the Achilles' heel of the American manufacturing industry. More than 300 million tons of hazardous waste are generated annually by about 14,000 installations in the United States. About 14.7 billion gallons of hazardous waste is disposed of in or on the land each year,

while around 500 million gallons are incinerated. Of the total quantity of hazardous waste generated, manufacturers account for 92 percent. It has been estimated that the chemical industry alone generates 71 percent of the manufacturer's total.[8] The Congressional Budget Office has estimated that the 1984 amendments to the Resource Conservation and Recovery Act could increase industrial compliance costs from between $4.2 billion and $5.8 billion in 1983 to between $8.4 billion and $11.2 billion in 1990, depending on the level of waste reduction achieved by industry.[9]

A variety of methods have been used over

the past 100 years to bury hazardous waste. Many of these burial sites now pose a threat to the health of nearby residents and, more broadly, to the nation's underground water supply (Figure 7.5). For example, recent studies by the state of California have shown that there are widespread threats to groundwater in California's Silicon Valley; Santa Clara County leads the nation in the number of sites on the National Priority List, most of which are associated with the electronics industry.[10] The U.S. Congress Office of Technology Assessment has recently projected that there are 10,000 sites nationwide that belong on the National Priority List of toxic waste dumps.[11]

In 1980, Congress appropriated $1.6 billion for a 5-year Superfund program. The original Superfund legislation viewed cleanup of hazardous waste sites as a relatively short-term program and anticipated that waste could be contained for several decades by methods such as building slurry walls and clay caps to eliminate diffusion of buried waste into subsurface

waters. After a few years of pursuing such methods, it is clear that they do not provide a solution to the problem of containment of waste in existing landfills; slurry walls leak and clay caps crack. It is also becoming increasingly clear that it will require decades to accomplish an adequate cleanup of hazardous waste sites nationwide. Thus, when the Superfund act was reauthorized in 1986, a significant focus was on the use of new technologies to decontaminate soil and groundwater and to provide for long-term containment of wastes (i.e., through encapsulation). The level of expenditure could be as high as $10 billion over the next 5 years.

When one takes the cost of industrial compliance with RCRA to handle currently generated wastes and adds the cost of Superfund to clean up the wastes of the past, it becomes obvious that there are strong incentives for technology development in the area of waste minimization and treatment, and many opportunities for research and employment for chemical engineers.

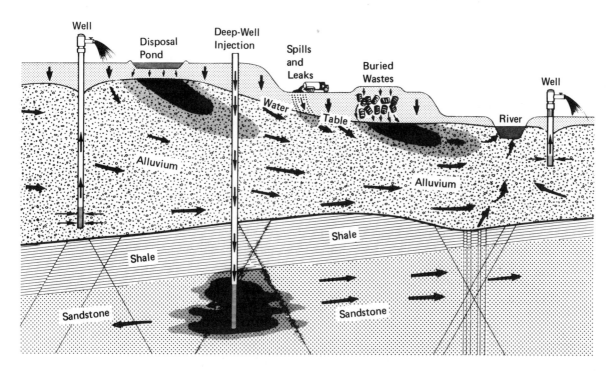

FIGURE 7.5 Past methods of waste disposal threaten water supplies today. Reprinted from *Opportunities in Chemistry*, National Academy Press, 1985.

DESIGN OF INHERENTLY SAFER AND LESS POLLUTING PLANTS AND PROCESSES

Few basic decisions affect hazard potential or have more of an impact on environment than the initial choice of technology. Thus, when designing chemical manufacturing processes, it is important to select sequences of chemical reactions that avoid the use of hazardous feedstocks and the generation of hazardous chemical intermediates. It is necessary to find reaction conditions tolerant of transient excursions in temperature, pressure, or concentration of chemicals and to use safe solvents when extracting reaction products during purification steps. Finally, it is important to minimize storage and in-process inventories of hazardous substances. The term "inherently safer plants" has been used to describe this approach.[12] One further consideration in the design of a new process or plant is whether it is going to generate polluting effluents or hazardous wastes. Good design should result in waste minimization in a manufacturing process or plant.

Traditional analyses of process economics might show that inherently safer and less polluting plants are less efficient in terms of energy or raw materials usage. Indeed, chemical plants have been designed in the past principally to maximize reliability, product quality, and profitability. Such issues as chronic emissions, waste disposal, and process safety have often been treated as secondary factors. It has become clear, however, that these considerations are as important as the others and must be addressed during the earliest design stages of the plant. This is in part due to a more realistic calculation of the economics of building and operating a plant. When potential savings from reduced accident frequency, avoidance of generating hazardous waste that must be disposed of, and decreased potential liability are taken into consideration, inherently safer and less polluting plants may prove to cost less overall to build and operate. And in any case, if the American public is not convinced that chemical plants are designed to be safe and environmentally benign, then the fact that they operate economically will be of little consequence to the public's

decision on whether to allow their construction and operation.

The chemical reaction pathways chosen for a manufacturing process profoundly influence chemical plant safety because they determine the nature and amounts of all substrates, products, and reagents and implicitly govern the design and operation of all hardware. The ultimate goal of chemical engineering research to provide inherently safer plants should be to elucidate the connection between process pathways and plant safety and to translate this connection into a quantitative form amenable to engineering design calculations. The array of chemicals, reactions, processes, and types of physical equipment used in industry is exceedingly diverse and constantly evolving. To effectively address the need for inherently safer plants, chemical engineering research must be focused on fundamental issues that span the entire range of processing activities, from elucidating detailed reaction mechanisms to understanding and predicting the gross response of coupled equipment. The goal of such fundamental research would be to develop the tools needed to define, discern, and assess the safety issues associated with a given process design and its alternatives.

New approaches to the design of commercial chemical syntheses should be pursued. A chemical synthesis tree graph with a high-value product at its apex, lower-value raw materials at the base, and reaction steps as nodes connecting all branches offers a basis for quantitative assessment of feasible and economic process alternatives. It could also serve to define the safety and environmental impact of a pathway and offer a basis for safe designs that produce minimal wastes. Tree graphs could be particularly helpful in evaluating the process safety implications of highly selective synthesis routes. Near the apex of the chemical synthesis tree graph materials that might be used in chemical reactions are of highest value. Overall raw material and energy costs are lowest for those pathways that use the most selective reactions to achieve the highest yield of the desired product. However, these selective reactions often require the handling and storage of more reactive, and hence more hazardous, chemicals.

For example, in the synthesis of Carbaryl at Bhopal, a process that required storage of large quantities of the reactive intermediate methyl isocyanate was used, rather than a less flexible and more expensive straight-through reaction scheme with a minimal inventory of methyl isocyanate. Developing the methodology of using process safety and environmental factors in synthesis tree graphs could provide a better framework for future plant design.

Most accidents in chemical plants occur when the plant is not operating at a steady state—for example, when it is starting up or shutting down or when a transient of temperature, pressure, or reactant concentration occurs. Fundamental research in non-steady-state process control and the management of process transients is therefore warranted. Design methodology poses a related research issue. It is obviously easier for the designer of a plant or individual reactor to envision how the equipment will operate during the normal production mode than to envision how it will operate under a host of potential scenarios that derive from process transients. The safety of chemical plants and reactors could be improved if designers had the means to envision the complete reaction topography and to assess the consequences of straying from normal operations. This would involve developing design tools that would incorporate chemical pathway information more systematically into classical engineering design methods for reactors and associated equipment.

COMBUSTION

Many of the environmental issues listed in Table 7.1 are intimately related to combustion. Combustion contributes significantly to emissions of pollutants into the environment, with effects ranging from those pertaining to indoor air pollution to those affecting global climate. For this reason, combustion has been singled out to illustrate the progress that can be made in resolving environmental issues through a sustained fundamental research program and to demonstrate the potential added benefits of continued in-depth study of the physical and chemical processes underlying combustion.

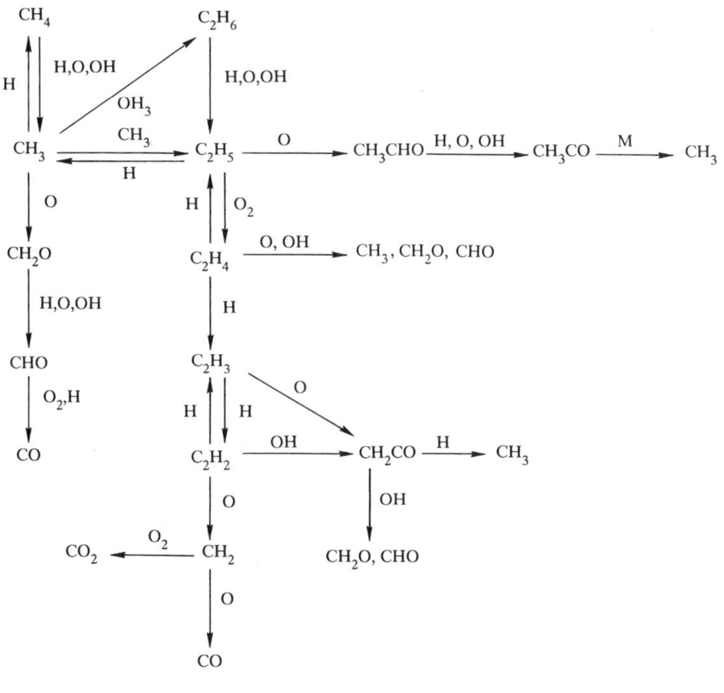

Hydrocarbons and Fuel-Bound Nitrogen

The burning of fuel in a practical combustion system, such as a power plant boiler or the cylinder of an internal combustion engine, is at first glance very simple: a mixture of hydrocarbon and air is ignited and burned to carbon dioxide and water. On closer examination this burning turns out to be one of the most complex processes in all engineering. For example, the combustion of the simplest hydrocarbon fuel, methane, involves more than 50 chemical reactions (Figure 7.6). During the past four decades, major progress has been made in developing a mechanistic understanding of the combustion of methane and C-2 hydrocarbons and their derivatives. Rate constants of many individ-

FIGURE 7.6 Mechanism of methane combustion.

ual free-radical reactions have been measured, and a good number of those not measured can be estimated from thermochemical kinetics and unimolecular reaction theory.

Major unknowns in the mechanism by which a hydrocarbon fuel burns concern the pyrosynthesis reactions that lead to the formation of polycyclic aromatic hydrocarbons (PAHs) and soot and the oxidation chemistry of atoms other than carbon and hydrogen (heteroatoms) in the fuel, particularly nitrogen, sulfur, and halogens.

Nitrogen oxide emissions from furnaces and boilers come mostly from oxidation of the nitrogen atoms in the fuel, whereas in internal combustion engines these emissions are derived largely from oxidation of atmospheric nitrogen. Burners of advanced design currently reduce the emissions of nitrogen oxides by a factor of 2 from uncontrolled combustion systems by staging the addition of oxygen to produce an initial fuel-rich regime in which the bound nitrogen is partially converted to N_2 (Figure 7.7). Potentially greater reduction in nitrogen oxides can be attained by adding hydrocarbons downstream of the fuel. This is called reburning (Figure 7.8). To determine the optimal sequence of air and fuel addition requires detailed knowledge of both the fuel nitrogen chemistry and the hydrocarbon chemistry.

The development of staged combustors for the control of nitrogen oxides is constrained partly by the formation of PAHs and soot (Figure 7.9). The PAHs are potential carcinogens whose biological activity depends strongly on their molecular structure. It is postulated that they are formed under locally fuel-rich conditions by the successive addition of C-2 through C-5 hydrocarbons to aromatic rings followed by ring closure. On the other hand, a staged combustor cannot be operated on too lean a fuel mixture because formation of nitrogen oxides is favored under this regime. Con-

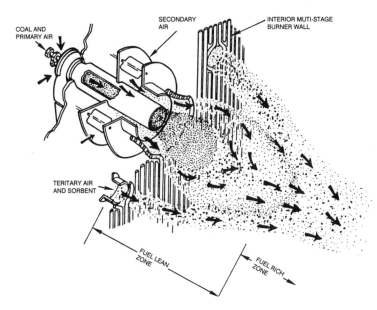

FIGURE 7.7 Schematic of a low-NO_x/SO_x pulverized coal burner. The addition of oxygen is staged to produce an initial fuel-rich zone in the burner that results in reduced emissions of nitrogen oxides.

tinued experimental and theoretical study of the combustion chemistry of higher molecular weight hydrocarbons should provide the understanding needed for the design of combustors in which the fuel/oxygen regime is selected to minimize emissions of PAHs and nitrogen oxides.

Soot

Combustion processes are a major source of particles emitted to the atmosphere. Particles formed in combustion systems fall roughly into

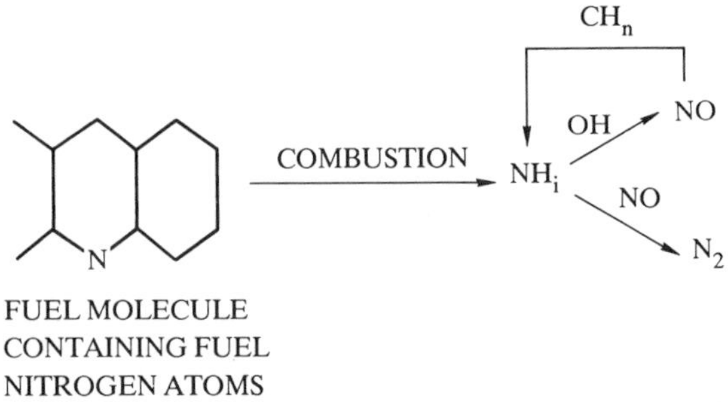

FUEL MOLECULE
CONTAINING FUEL
NITROGEN ATOMS

FIGURE 7.8 NO_x control in combusion by reburning. Addition of hydrocarbons (CH_n) late in the combustion process leads to the reduction of nitrous oxide (NO) to nitrogen gas (N_2).

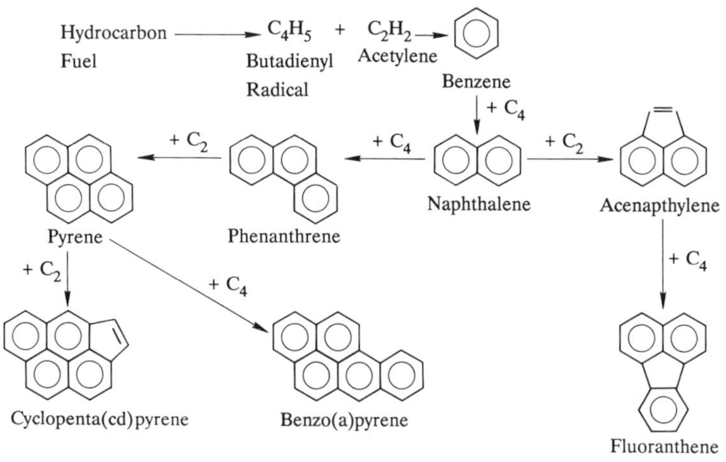

FIGURE 7.9 Mechanism of formation of polycyclic aromatic hydrocarbons (PAHs) during combustion.

compounds are formed (Figure 7.10). An understanding of the mechanism of soot formation has become more important because soot hampers the use of such important technologies as staged combustion and diesel engines. In addition, the sooting tendency of aromatic compounds is higher than that of aliphatics, and the aromatic content of fuels is expected to increase in the future as petroleum resource availability forces refiners to use fuel feedstocks with lower ratios of hydrogen to carbon.

Soot is objectionable not only because of its opacity but also because soot particles are carriers of toxic compounds. When combustion products cool, soot particles provide condensation sites for hydrocarbon vapors, particularly PAHs. Soot particles are agglomerates of small, roughly spherical units. The small vary in diameter from 0.005 to 0.2 μm, with most in the range of 0.01 to 0.05 μm, while the size and morphology of the clusters can range from aggregates of several particles to large contrails several micrometers in diameter and hundreds of micrometers in length. Soot particles are not pure carbon. The atomic ratio of hydrogen to carbon decreases from around 1.0 at the point

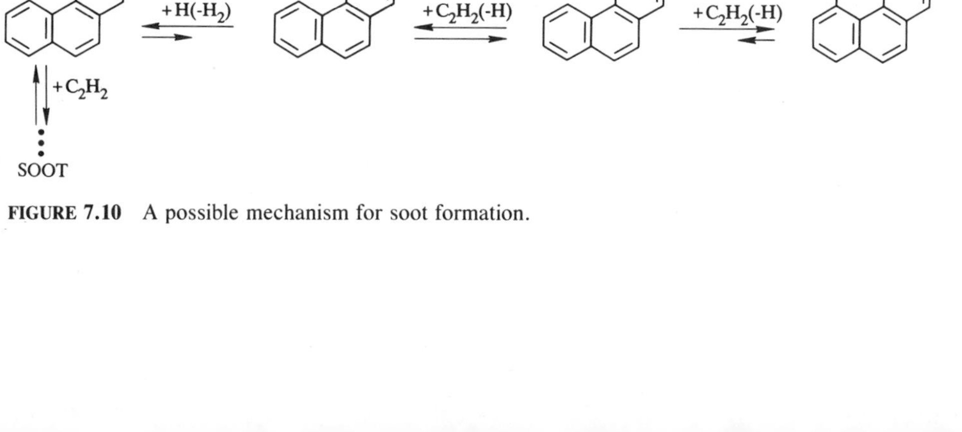

two categories. The first, referred to as soot, consists of carbonaceous particles formed by pyrolysis of the fuel molecules. The second, referred to as ash, is composed of particles derived from noncombustible constituents in the fuel and from heteroatoms in the organic structure of the fuel.

Soot can be produced in the combustion of gaseous fuels and from the volatilized components of liquid or solid fuels. Soot formation is a complex process involving the chemistry of fuel destruction under fuel-rich conditions where hundreds of aromatics and other intermediate

FIGURE 7.10 A possible mechanism for soot formation.

of first formation in the flame to 0.1 to 0.2 in the cooled exhaust.

There is considerable uncertainty about the mechanisms of soot particle inception and growth. To evaluate potential measures to suppress soot formation or to accelerate its postflame oxidation, it is important to understand how fuel structure and combustion conditions influence the physical and chemical nature of the soot particle. Particle inception in the flame occurs by some sort of nucleation process not yet understood. Once formed, the soot particles grow by surface reactions of hydrocarbon radicals. It is generally accepted that acetylene is an important intermediate for growth of soot particles, but the precise pathway by which the acetylene contributes to soot growth is not known. Although significant progress has been made in simulating the dynamics of multicomponent aerosols, basic information on the fundamental chemistry and physics of soot formation and growth is needed to enable researchers to predict the size and chemical composition of soot particles as a function of fuel type and combustion conditions.

Ash

The United States uses a disproportionate amount of gas and oil compared to coal (Figure 7.11). It has long been known that the country must increase the fraction of its energy that is derived from coal if it is to be independent of foreign energy sources. The major obstacles to the wider use of coal are environmental. The mineral content of U.S. coals averages about 10 percent by weight, and the sulfur content varies widely around an average of approximately 2.5 percent. The minerals in coal lead to the formation of ash particles, and the sulfur leads to sulfur dioxide in the flue gas.

Ash particles produced in coal combustion are controlled by passing the flue gases through electrostatic precipitators. Since most of the mass of particulate matter is removed by these devices, ash received relatively little attention as an air pollutant until it was shown that the concentrations of many toxic species in the ash particles increase as particle size decreases. Particle removal techniques become less effec-

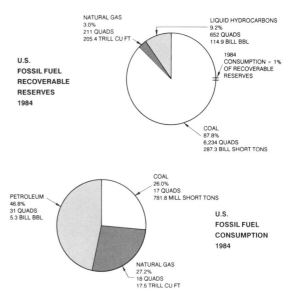

FIGURE 7.11   U.S. fossil fuel reserves and consumption in 1984. Courtesy, Department of Energy.

tive as particle size decreases to the 0.1–0.5 μm range, so that particles in this size range that escape contain disproportionately high concentrations of toxic substances.

The processes that govern the formation of ash particles are complex and only partially understood (Figure 7.12). The mineral matter in pulverized coal is distributed in various forms; some is essentially carbon-free and is designated as extraneous; some occurs as mineral inclusions, typically 2–5 μm in size, dispersed in the coal matrix; and some is atomically dispersed in the coal either as cations on carboxylic acid side chains or in porphyrin-type structures. The behavior of the mineral matter during combustion depends strongly on the chemical and physical state of the mineral inclusions.

During combustion the mineral inclusions decompose and fuse. Most of the mineral matter adheres to the char surface, but some is released as micrometer-sized particles. As the char surface recedes during burning, the ash inclusions are drawn together and coalesce to form larger ash particles. Char fragments are released and take with them ash inclusions and ash adhered to the surface. As each char fragment burns out, an ash particle is produced of a size and composition determined by the evolution of the char pore structure during combustion.

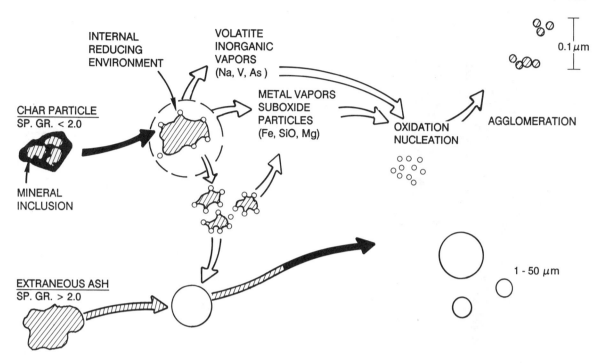

FIGURE 7.12 Fly ash formation during coal combustion.

The high temperatures of coal char oxidation lead to a partial vaporization of the mineral or ash inclusions. Compounds of the alkali metals, the alkaline earth metals, silicon, and iron are volatilized during char combustion. The volatilization of silicon, magnesium, calcium, and iron can be greatly enhanced by reduction of their refractory oxides to more volatile forms (e.g., metal suboxides or elemental metals) in the locally reducing environment of the coal particle. The volatilized suboxides and elemental metals are then reoxidized in the boundary layer around the burning particle, where they subsequently nucleate to form a submicron aerosol.

There is a general understanding that the size of ash particles produced during coal combustion decreases with decreasing coal particle size and with decreasing mineral content of the parent coal particles. There are, however, no fundamental models that allow the researchers to predict the change in the size of ash particles when coal is finely ground or beneficiated or how ash size is affected by combustion conditions.

Sulfur Oxides

Current strategies for reducing sulfur oxide emissions from coal-fired combustors are based on the addition of calcium sorbents that retain the sulfur either as a sulfide in a reducing slagging combustor or as a sulfate formed in entrained flow in a pulverized coal boiler or fluidized bed (Figure 7.13). The utilization of the calcium is currently as low as 20 percent, which results in a large volume of spent sorbent. The problems preventing better sorbent utilization are the sintering of pores at high temperatures near the flame zone, the low diffusivity of sulfur dioxide through the layer of calcium sulfate that forms on grains of the calcium oxide sorbent, and pore plugging. There is opportunity for major innovation in the design of sorbents for sulfur capture in combustors by tailoring their physical and chemical properties. The key characteristics of an ideal sorbent are large surface area, mechanical strength, and fast and complete utilization. Used sorbent should be regenerable or usable as a by-product.

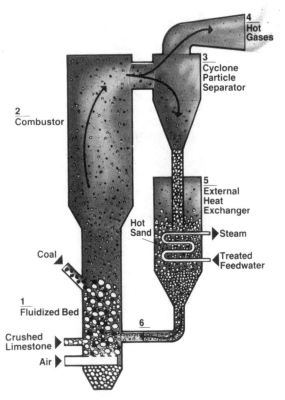

FIGURE 7.13 Fluidized bed combustion. In fluidized bed combustion, air is blown into a bed of burning coal, limestone, ash, and gravel, swirling the mixture like a fluid and greatly improving combustion (1). The hot stream of gases generated carries small particles up the combustor (2). The particles of ash, gypsum, and limestone (called "sand") in the gas stream are separated and collected by large cyclone separators (3). The hot gases (4) are ducted to preheat boiler water. The hot sand transfers heat to boiler tubes (5), generating steam. Cooled sand particles (6) are recycled to the combustor. Courtesy, E.I. du Pont de Nemours and Company, Inc.

Fires and Explosions

Fires and explosions cause major property loss within the chemical process industry; more significantly, they account for an annual loss in this country of thousands of lives and the destruction of billions of dollars of property. The chemical engineer can contribute to the solution of the overall fire problem by providing means of estimating the flammability, flame propagation rates, and products of incomplete combustion for the increasing diversity of industrial and manufacturing materials, including polymer and ceramic composites. The problem

is complicated by the strong link between flame propagation and the configuration of and ventilation in different enclosures, with high-rise atriums in buildings being of special concern. Examples of the pressing problems to which the chemical engineer can contribute follow.

At present there is no small-scale test for predicting whether or how fast a fire will spread on a wall made of flammable or semiflammable (fire-retardant) material. The principal elements of the problem include pyrolysis of solids; char-layer buildup; buoyant, convective, turbulent-boundary-layer heat transfer; soot formation in the flame; radiative emission from the sooty flame; and the transient nature of the process (char buildup, fuel burnout, preheating of areas not yet ignited). Efforts are needed to develop computer models for these effects and to develop appropriate small-scale tests.

Most fire deaths are caused by smoke inhalation rather than by burns. Buildings now contain many synthetic polymeric materials that can burn to yield such toxic compounds as hydrogen cyanide and hydrogen chloride in addition to common combustion products such as carbon monoxide. For this reason, consideration is being given to banning certain materials, at least in public buildings. Realistic hazard analyses for materials would be facilitated by computer models that could interrelate such significant factors as the identity and amount of toxic products formed by combustion of these materials, the rate at which the materials burn, and the ease of ignition and smoke-forming tendency of the materials. Furthermore, as combustion products are transported away from the flame (e.g., down a corridor), smoke particles agglomerate and hydrogen chloride undergoes mass transfer to and adsorption on a variety of surfaces. Any interdisciplinary effort to understand the hazards of fires involving synthetic materials would benefit from chemical engineering research expertise in reaction modeling, chemical kinetics, and heat and mass transfer.

A small fire in a computer room, a telephone exchange, or an assembly plant for communication satellites can cause enormous damage because of minute amounts of corrosion on circuit elements. Furthermore, if either water or a halogenated agent is used to control the

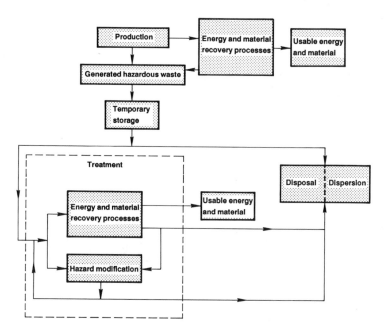

FIGURE 7.14 Strategies for dealing with continued generation of hazardous waste. Management paths for hazardous waste include temporary storage, treatment, disposal, and dispersion. Courtesy, Office of Technology Assessment.

fire, the agent itself or its decomposition products may cause damage to sensitive materials or devices. As automated and robotic systems become more prevalent in manufacturing plants, vulnerability of plants to small fires will increase. Chemical engineering research relevant to this challenge includes the development of sensors for more sophisticated fire detection, the design and development of materials and techniques for encapsulation of sensitive device elements, research on surfaces and interfaces to facilitate more effective equipment salvage, and research to develop a better understanding of corrosion mechanisms so that optimal strategies for fighting small fires can be developed.

One of the great hazards in a chemical plant is the potential for any deflagration (a fire in which the flame front moves through the combustible mix at subsonic velocities) to turn into a detonation, in which the flame front propagates at supersonic velocities, generating a blast wave. Considerable basic research has been conducted to understand this transition in the combustion of hydrogen and significant progress has been made. Extrapolating this understanding to more complex compounds and to mixtures of the chemicals found in chemical plants is a challenging problem.

HAZARDOUS WASTE MANAGEMENT

There are three distinct problems in hazardous waste management: (1) reduction in the generation of waste, already mentioned in the section "Design of Inherently Safer and Less Polluting Plants and Processes"; (2) disposal of generated waste (Figure 7.14); and (3) remediation of old, abandoned waste disposal sites. The problems of handling and disposing of radioactive waste are largely the concern of nuclear engineers, often working with chemical engineers to develop separation and encapsulation technologies for radioactive nuclides, and are not discussed here. The most basic way to deal with the continuing generation of hazardous waste is to accumulate, encapsulate, and store only as a temporary measure and to develop new approaches to reduce the volumes generated and to concentrate hazardous components or convert them into nonhazardous materials. Abandoned waste sites are remediated by cleaning them up or containing them before they contaminate groundwater supplies. The establishment of priorities for site cleanup and the development of appropriate detoxification technologies require an understanding of the processes by which the waste can migrate or be transformed in the natural environment.

The development of a fundamental understanding of the behavior of toxic chemicals in atmospheric, soil, and aquatic environments (Figure 7.15) and of possible mechanisms for destroying toxic chemicals has lagged far behind the rapidly unfolding problems surrounding disposal. Nevertheless, the ability of American manufacturing industries to remain internationally competitive depends on this. Engineers in

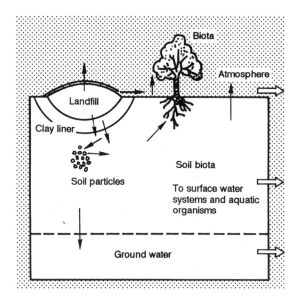

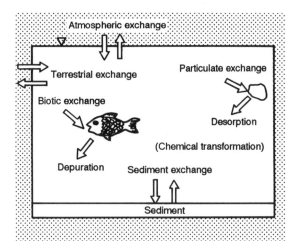

FIGURE 7.15 Transport and transformation of toxic chemicals in soil environments (left) and water environments (right). Courtesy, Office of Technology Assessment.

industry did not anticipate that the problems associated with hazardous waste disposal would emerge so rapidly or that destruction and disposal processes would be so difficult to develop. A major effort must be mounted to conduct advanced research and to educate engineers to solve the problems associated with the disposal and environmental behavior of toxic chemicals.

Detoxification of Currently Generated Waste

Many technologies have been proposed for detoxifying waste by processes that destroy chemical bonds: pyrolytic; biological; and catalyzed and uncatalyzed reactions with oxygen, hydrogen, and ozone. The following sections deal only with research opportunities in the areas of thermal destruction, biodegradation, separation processes, and wet oxidation.

Thermal Destruction

Most organic molecules are not stable at temperatures above 400°C for any significant length of time. Therefore, many processes for detoxifying waste will heat the toxic compounds to temperatures at which they will rapidly de-

compose. Heating methods include resistive electrical heating, the use of radio frequencies or microwaves, radiative heating, and the use of hot combustion products as a heat source. Heating can be done in the absence of oxygen, in which case the process is known as pyrolysis. Pyrolysis yields different products than does combustion of waste in the presence of oxygen.

The most important current technique for the thermal destruction of waste is incineration, where the energy required for destruction is provided by oxidation of the waste, sometimes supplemented with a fossil fuel. The major question about all thermal destruction techniques is whether products from the process—either traces of unreacted parent compound or compounds synthesized from the parent compound at high temperature—will pose a health hazard.

Concerns have been expressed about incineration on land and in the water. EPA's Science Advisory Board, in a 1984 report entitled *Incineration of Hazardous Liquid Waste*, stated, "The concept of destruction efficiency used by the EPA was found to be incomplete and not useful for subsequent exposure assessments."[13] It was recommended that the emissions and effluents of hazardous waste incinerators be analyzed in such a way that the identity, quantity, and physical characteristics of the chemicals released into the environment could be estimated. The International Maritime Organization Scientific Group on Ocean Dumping, convened in London in March 1985, was unable to reach consensus on the following questions:

- What is the relationship between destruction and combustion efficiencies over a wide range of operation conditions?
- What sampling procedures should be used to obtain a gas sample representative of the entire stack?
- What methodology should be used for collecting particulate matter in the stack?
- What new organic compounds can be synthesized during the incineration process?

Factors that influence the destruction efficiency in incineration include

- local temperatures and gas composition,
- residence time,
- extent of atomization of liquid wastes,
- dispersion of solid wastes,
- fluctuations in the waste stream composition and heating value,
- combustion aerodynamics, and
- turbulent mixing rates.

Mere destruction of the original hazardous material is not, however, an adequate measure of the performance of an incinerator. Products of incomplete combustion can be as toxic as, or even more toxic than, the materials from which they evolve. Indeed, highly mutagenic PAHs are readily generated along with soot in fuel-rich regions of most hydrocarbon flames. Formation of dioxins in the combustion of chlorinated hydrocarbons has also been reported. We need to understand the entire sequence of reactions involved in incineration in order to assess the effectiveness and risks of hazardous waste incineration.

The routine monitoring of every hazardous constituent of the effluent gases of operating incinerators is not now possible. EPA has established procedures to characterize incinerator performance in terms of the destruction of selected components of the anticipated waste stream. These compounds, labeled principal organic hazardous components (POHCs), are currently ranked on the basis of their difficulty of incineration and their concentration in the anticipated waste stream. The destruction efficiency is expressed in terms of elimination of the test species, with greater than 99.99 percent removal typically judged acceptable provided that toxic by-products are not generated in the process.

The effectiveness of incineration has most commonly been estimated from the heating value of the fuel, a parameter that has little to do with the rate or mechanism of destruction. Alternative ways to assess the effectiveness of incineration destruction of various constituents of a hazardous waste stream have been proposed, such as assessment methods based on the kinetics of thermal decomposition of the constituents or on the susceptibility of individual constituents to free-radical attack. Laboratory studies of waste incineration have demonstrated that no single ranking procedure is appropriate for all incinerator conditions. For example, acceptably low levels of some test compounds, such as methylene chloride, have proved difficult to achieve because these compounds are formed in the flame from other chemical species.

Rather than focus on specific incineration technologies, one must address the fundamental physical and chemical processes common to many of the possible incineration systems through studies of (1) reaction kinetics of selected waste materials and (2) behavior of waste solutions, slurries, and solids in the incineration environment.

The combustion chemistry of methane and C-2 hydrocarbons is reasonably well understood. Progress is being made in addressing the pyrosynthesis reactions that lead to the formation of toxic PAHs. Much of the literature on combustion, though, is devoted to the flame zone, where heat release rates and free-radical concentrations are high. A key problem in incineration chemistry involves understanding the late stages of degradation of waste materials, where temperatures and free-radical concentrations are lower than in the flame zone. Moreover, it has long been known that the introduction of halogen atoms into a flame can interfere with the combustion process by removing free radicals. The effect of such reactions on the incineration of hazardous substances containing halogen atoms needs to be determined. We are concerned both with the destruction of the original compounds and with the production of trace quantities of other hazardous species during the reaction. Thermal pyrolysis and reactions of the waste with common radicals such as OH, O, H, and the halogens are most important in the flame environment. Both reducing and oxidizing atmospheres are encountered in turbulent diffusion flames; therefore, an understanding is needed of the chemistry over the entire range of combustion stoichiometries.

Studies of the incineration of liquid and solid wastes must determine the rates at which hazardous compounds are released into the vapor phase or are transformed in the condensed phase, particularly when the hazardous materials make up a small fraction of the liquid burned. We must be particularly concerned with understanding the effects of the major composition and property variations that might be encountered in waste incinerator operations—for example, fluctuations in heating value and water content, as well as phase separations. Evidence of the importance of variations in waste properties on incinerator performance has been demonstrated by the observation of major surges in emissions from rotary-kiln incinerators as a consequence of the rapid release of volatiles during the feeding of unstable materials into the incinerator.

Biodegradation

Recent developments in molecular biology, particularly recombinant DNA technology, offer many new concepts for hazardous waste treatment that were unthinkable a decade ago. For example, the insertion of foreign genes from one microorganism into another has become relatively routine. Progress in achieving high expression of a foreign inserted gene has also been impressive. A combination of molecular biology and chemical engineering could lead to the design of new processes for waste treatment.

The controlled use of biological systems or their products to bring about chemical or physical change is particularly attractive when dealing with dilute waste streams. Biological systems thrive in dilute aqueous media, where they can effectively degrade organic pollutants, absorb heavy metal ions, or change the valence state of heavy metal ions (Table 7.2). Microorganisms and biological agents can carry out reactions with great chemical specificity and efficiency, and genetic engineering provides means for developing strains of organisms and classes of enzymes with nearly unlimited capabilities for effecting desired chemical changes. In addition, many microbial systems have high affinity for metal ions, and metal ions are often moved from an aqueous solution into the cell through active transport. Accordingly, such reactions as the biological reduction of a heavy metal ion can be carried out at relatively fast rates, although at millimolar concentrations.

There are significant research opportunities for chemical engineers in the design and opti-

---

TABLE 7.2 Examples of Biodegradation for Waste Management

Industry	Effluent Stream	Major Contaminants Removed by Biodegradation
Steel	Coke-oven gas scrubbing operation	Ammonia, sulfides, cyanides, phenols
Petroleum refining	Primary distillation process	Sludges containing hydrocarbons
Organic chemical manufacture	Intermediate organic chemicals and by-products	Phenols, halogenated hydrocarbons, polymers, tars, cyanide, sulfated hydrocarbons, ammonium compounds
Pharmaceutical manufacture	Recovery and purification solvent streams	Alcohols, ketones, benzene, xylene, toluene, organic residues
Pulp and paper	Washing operations	Phenols, organic sulfur compounds, oils, lignins, cellulose
Textile	Wash waters, deep discharges	Dyes, surfactants, solvents

SOURCE: Office of Technology Assessment.[14]

mization of bioreactors for dilute waste stream treatment, including the design of efficient contactors, the use of immobilized cells in reactors, and the elucidation of mass transport processes and reaction kinetics. Related research opportunities for chemical engineers include the formulation of biocatalysts, the development of bioseparations, and the use of chemical engineering expertise in process control and optimization to better understand the behavior of large microbial populations.

The promise of biological treatment of heavy metal ions has already been illustrated by strains of microorganisms that tolerate mercury, chromium, and nickel—heavy metals that are generally toxic to microorganisms. Tolerant microorganisms have been isolated through classical adaptation and strain-selection studies rather than by recombinant DNA techniques. Mercury-tolerant microorganisms have been shown to possess an enzyme that is not present in nonresistant strains. This enzyme, mercuric reductase, is able to catalyze the reduction of mercury(II) ions to metallic mercury. Since the mercury(II) ion is the toxic species and the insoluble metal is chemically inactive, the microorganism is able to detoxify a solution that contains mercury(II) ions.

Microorganisms hold tremendous promise for improvements in the treatment of hazardous waste, but genetically altered microorganisms present both regulatory bodies and industry the complex task of identifying, managing, and controlling their use. While organisms that have been highly modified by classical strain selection have been used safely in industry for years, there is a critical lack of data to either support or to allay concerns about the release into the open environment of organisms that have been modified by recombinant DNA techniques. For example, to understand the potential consequences of the release of genetically engineered organisms, it is also necessary to know

- alternative approaches to meeting the need,
- alternative systems for handling the organism,
- how much material will be involved,
- how the organisms will move or be transported,
- what chemical substances will be produced,

- ecosystem interactions,
- exposure pathways, and
- probable short- and long-term impacts.

A substantial base of scientific information, monitoring methods, and predictive models is required. Chemical engineers can assist biologists in developing this base. For example, the chemical engineering tools used to analyze chemical processes in industrial reactors would be useful in analyzing a situation where genetically engineered microorganisms were released into an underground waste site, with the earth itself being the chemical reactor.

Separation Processes

A large fraction of the hazardous waste generated in industry is in the form of dilute aqueous solutions. The special challenges of separation in highly dilute solutions may be met by the development of new, possibly liquid-filled, membranes; by processes involving selective concentration of toxic chemicals on the surfaces of particles; or by the use of reversed micelles.

Liquid-filled, porous, hollow-fiber membranes hold promise for improving the efficiency and economics of extraction processes. In conventional liquid-liquid extraction, process design, hardware, and economics are dictated primarily by the relative densities of the two liquid phases. Energy must be expended to create a large surface area for diffusion to take place, while contact time may be shorter than desired because of high relative velocities of the two phases. Hollow fibers containing pores filled with the extracting agent permit the waste and stripping fluids to flow in a countercurrent fashion on opposite sides of the membrane. In this way, a high interfacial area can be maintained, regardless of the relative flow rate of fluid to extracting agent. In addition, the extracting agent can be renewed by continuous desorption into the stripping fluid. This technique is but one example of many new processes evolving in the field of membrane separations.

Another process for the separation of toxic chemicals from waste streams species involves adsorption from solution onto particles, followed by sedimentation to remove the toxic-laden particles. Solutes bound to the surface of aqueous particles may participate in oxidation-reduction reactions with the particles, undergo chemical transformations in which the particle surface serves as a catalyst, or participate in heterogeneous photochemical processes. The design of effective engineering processes for the treatment of water supplies to remove toxic compounds by adsorption/reaction/particle-removal sequences demands fundamental data on the kinetics of the individual steps and the incorporation of the data into process models. A major challenge is to describe all relevant chemical influences on the efficiency of removal of specific toxic compounds. Among these are the physical and chemical properties of the absorbing particle surface, the alteration of these properties by reactions or dissolution-precipitation processes, and the stability of aqueous particles to coagulation. In contrast to the chemical conditions of conventional municipal water and wastewater processing, the conditions selected or imposed by the special circumstances for control of hazardous substances may include extremes of pH, redox potential, ionic content, and organic content. These factors may become critical in the design of optimal processes combining adsorption, reaction, and coagulation steps.

Recent development of the use of reversed micelles (aqueous surfactant aggregates in organic solvents) to solubilize significant quantities of nonpolar materials within their polar cores can be exploited in the development of new concepts for the continuous selective concentration and recovery of heavy metal ions from dilute aqueous streams. The ability of reversed micelle solutions to extract proteins and amino acids selectively from aqueous media has been recently demonstrated; the results indicate that strong electrostatic interactions are the primary basis for selectivity. The high charge-to-surface ratio of the valuable heavy metal ions suggests that they too should be extractable from dilute aqueous solutions.

The potential of reversed micelles needs to be evaluated by theoretical analysis of the metal ion distribution within micelles, by evaluation of the free energy of the solvated ions in the reversed micelle organic solution and the bulk aqueous water, and by the experimental char-acterization of reversed micelles by small-angle neutron and x-ray scattering.

Wet Oxidation

Incineration achieves high destruction efficiencies by fast free-radical reactions in the presence of water vapor at high temperatures (1,500–2,300°C) and 1 atm. Waste can also be destroyed by oxidation at much lower temperatures by operating at high pressures, including conditions above the critical point for water. For example, high destruction efficiencies (greater than 99.99 percent) of toxic organic compounds can be achieved in 2 seconds at moderate temperatures (1,000–1,200°C) at 250 atm. The lower reaction temperature permits the destruction of waste of much lower heating value than can be incinerated, at least without the use of auxiliary fuels. The chemistry of such reactions at high pressures and moderate temperatures needs to be further elucidated before wet oxidation processes can be more widely used in hazardous waste management.

Remediation of Toxic Waste Sites

Only two processes, high-temperature pyrolysis and mobile incineration, have proved effective for soil decontamination and are considered to be commercially viable. Both involve heating the contaminated soil to a high temperature, which is costly in terms of energy use and materials handling. There are substantial opportunities for innovation and development of processes for the separation of contaminants from soils and the in-situ treatment of contaminated soils. Examples of each are given in the following subsections.

Separation Processes

One generic problem in site remediation is the removal or deactivation of small quantities of toxic organics from highly porous and surface-active media such as soil. Alternative processes to pyrolysis and high-temperature oxidation of soil, such as thermal desorption, steam stripping, and supercritical extraction, require less energy and thus should be investigated further. Fundamental research on the nature of the adsorbed state of organics in soil could have as a significant payoff the identification of alternative process paths. Basic measurements of desorption kinetics and pore diffusion in classified fractions of soil components (e.g., clays, silts, and sand) can provide the basis for developing accurate models of such processes as soil desorption and migration of contaminated plumes. This information could be used to determine the conditions necessary for thermal desorption and steam stripping.

Extraction with supercritical fluids, such as carbon dioxide or methanol in carbon dioxide, offers the potential for combining the high mass-transfer coefficients of gases with the moderately high absorption capacities of liquid solvents. In addition, the solubility characteristics are highly sensitive to relatively small changes in temperature and pressure. Thus, the contaminants can be recovered from the supercritical fluid after extraction and the supercritical fluid recycled at moderate cost. The method is being applied to tertiary oil recovery by the petroleum industry, a process somewhat akin to the removal of organics from soil. Fundamental research on the solubilities of organic compounds in supercritical fluids would expedite the evaluation and application of this promising technology.

Biodegradation

The use of biodegradation for the treatment of dilute waste streams has already been discussed; it also has potential for in-situ treatment. The critical need is to learn how to select and control microorganisms in a soil environment to achieve the desired degradation of organics.

Monitoring

One of the most important elements in the remediation of existing waste sites is early detection and action. As an example, the cost of cleanup at Stringfellow, California, increased from an estimated $3.4 million to $65 million because of pollutant dispersal during a decade of inaction after the first identification of the problem. The opportunities for innovative sampling strategies responsive to this need are discussed in the following section.

BEHAVIOR OF EFFLUENTS IN THE ENVIRONMENT

It has been recognized for some time that fluids in motion, such as the atmosphere or the ocean, disperse added materials. This property has been exploited by engineers in a variety of ways, such as the use of smoke stacks for boiler furnaces and ocean outfalls for the release of treated wastewaters. It is now known that dilution is seldom the solution to an environmental problem; the dispersed pollutants may accumulate to undesirable levels in certain niches in an ecosystem, be transformed by biological and photochemical processes to other pollutants, or have unanticipated health or ecological effects even at highly dilute concentrations. It is therefore necessary to understand the transport and transformation of chemicals in the natural environment and through the trophic chain culminating in man.

The Atmospheric Environment

Over the last two decades, significant progress has been made in understanding the mechanisms of transport and transformation of pollutants in the atmosphere. Mathematical models have been developed to describe the spatial and temporal distributions of sulfur dioxide, carbon monoxide, nitrogen oxides, hydrocarbons, and ozone. These models now serve as the backbone for the development of state plans for implementing the 1977 Clean Air Act amendments. Regionwide air pollution and acid rain are current subjects of intensive mathematical modeling efforts. But in spite of the strides that have been taken, a number of important research problems remain in understanding the behavior of atmospheric contaminants.

Organic compounds constitute about 25–30 percent of the fine aerosol mass (the mass contained in particles smaller than 2.5 μm diameter) in urban areas. They are of considerable interest because some of them, such as PAHs, are either suspected carcinogens or known mutagens. Still, little headway has been made toward engineering their systematic reduction in the atmosphere.

The problem is complex because many different sources contribute to atmospheric loadings of organic compounds. Not only do toxic waste incinerators have to be considered, but so do more than 50 classes of mobile and stationary combustion sources and industrial processes that release small amounts of toxic organics mixed with other exhausts. In addition, reliable aerosol source samples of PAHs and their oxygenated or nitrated derivatives are difficult to collect because these compounds are present in both gas and aerosol phases. Special stack-sampling equipment must be designed to acquire meaningful samples. Emissions undergo transport and chemical transformation in the atmosphere. For example, mutagenic nitro compounds can be created by the reaction of PAHs with HNO_3, NO_2, or N_2O_5. One way to analyze these atmospheric transformations is to compare the chemical composition of primary source effluents with that of ambient aerosol samples. However, source and ambient samplings now vary in methodology and analysis so that differences between them may be due to laboratory procedures. A comprehensive study, in which source and ambient measurements are made and analyzed the same ways, is needed. Source emission data could then be correlated with atmospheric transport calculations, and the relative importance of source contributions to ambient organics could be identified.

When spills and releases of hazardous gases or liquids occur, the concentration of the hazardous material in the vicinity of the release is often the greatest concern, since potential health effects on those nearby will be determined by the concentration of the substance at the time of the acute exposure. There are many models of routine *continuous* discharges (e.g., discharges arising from leaky valves in chemical plants), but these cannot be applied to single episodic events. Research on the ambient behavior of short-term environmental releases and the development of models for concentration profiles in episodic releases are crucial if we are to plan appropriate safety and abatement measures.

Because most people spend the majority of their time indoors, the quality of the indoor atmospheric environment is now receiving greater attention from researchers and regulators. There has been a reported increase in both the concentration and diversity of pollutants in indoor environments; formaldehyde, nitrogen dioxide, carbon monoxide, and a diverse range of organic compounds have been identified. It is not certain whether this should be attributed to the use of new building materials and to changes in building ventilation resulting from increased insulation or to the use of more sophisticated analytical techniques. Since the principal indoor air pollutants are known or suspected to adversely affect health (Figure 7.16), there is a need to engineer systems that can reduce their generation. Chemical engineers can assist in developing such systems, including

• home heating and cooking burners that minimize the generation of oxides of nitrogen;

• improved heat transfer devices that will allow for air exchange with the outside environment while avoiding excessive loss of heat; and

• resins, binders, coatings, and glues for building materials that do not emit hazardous compounds, such as formaldehyde.

Finally, there is a need for simple instrumentation that can be used to quantify occupant exposure to air pollution.

The Aquatic and Soil Environments

Disturbingly little is known about the mechanisms of groundwater contamination, including not only those for transport and dispersion but also those for chemical transformation. Underground pollutant transport is often represented with rather simplistic plume models, in much the same way as traditionally done for the atmosphere. These models do not take into account the fact that, for the underground transport process alone, the detailed mechanisms of flow through inhomogeneous porous media represent a major source of added complexity that cannot be ignored (Figure 7.17). Chemical engineering expertise in petroleum reservoir modeling can be applied to this area.

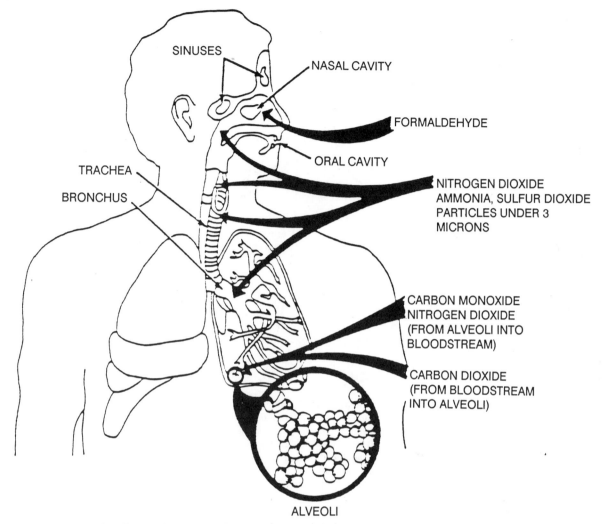

SINUSES

NASAL CAVITY

FORMALDEHYDE

ORAL CAVITY

TRACHEA

BRONCHUS

NITROGEN DIOXIDE
AMMONIA, SULFUR DIOXIDE
PARTICLES UNDER 3
MICRONS

CARBON MONOXIDE
NITROGEN DIOXIDE
(FROM ALVEOLI INTO
BLOODSTREAM)

CARBON DIOXIDE
(FROM BLOODSTREAM
INTO ALVEOLI)

ALVEOLI

FIGURE 7.16 Health effects of indoor air pollution. Gaseous and particulate contaminants frequently found in indoor air pollution affect different parts of the respiratory system. Some, such as carbon monoxide and nitrogen dioxide, move from the lungs into the bloodstream.

Models of chemical reactions of trace pollutants in groundwater must be based on experimental analysis of the kinetics of possible pollutant interactions with earth materials, much the same as smog chamber studies considered atmospheric photochemistry. Fundamental research could determine the surface chemistry of soil components and processes such as adsorption and desorption, pore diffusion, and biodegradation of contaminants. Hydrodynamic pollutant transport models should be upgraded to take into account chemical reactions at surfaces.

Considerable work has been done on the behavior of pollutant species at air-water and air-soil interfaces. For example, wet and dry deposition measurements of various gaseous and particulate species have been made over a wide range of atmospheric and land-cover conditions. Still, the problem is of such complexity that species-dependent and particle-size-dependent rates of transfer from the atmosphere to water and soil surfaces are not completely understood. There is much to be learned about pollutant transfer at water-soil interfaces. Concern about groundwater contamination by min-

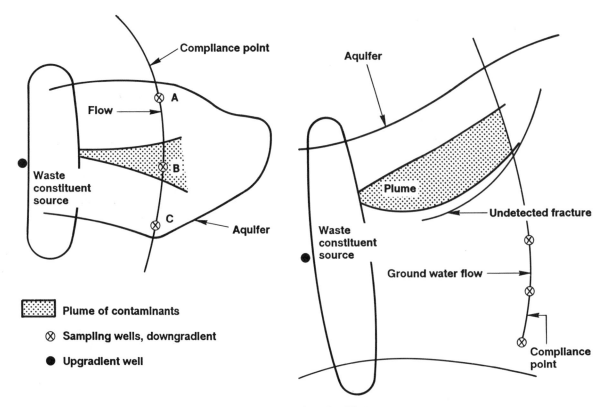

FIGURE 7.17 Oversimplified and more realistic views of plume migration in underground water. Plume migration is affected by inhomogeneities in the aquifer. The diagram on the right shows that gravitational influences or fractures on the aquifer might cause plumes to flow in directions different from the direction of groundwater flow. Courtesy, Office of Technology Assessment.

eral processing leachates, by rainwater leaching of landfills, and by runoff of contaminated surface water has heightened the opportunities for further work in this area.

Ambient Monitoring

Advances in understanding the transport and fate of chemicals in the environment will depend on substantial improvements in measurement capabilities. Attention should be directed toward instrumental techniques that can determine the oxidation state of inorganic species, which often has a marked influence on reactivity, transport properties, and toxicity of the ion. Free radicals and other highly reactive trace species play important roles in the chemistry of the environment. Detection of these species

requires rugged and reliable instrumentation that can be transported and used in the field. Remote sensing technology should be explored as a means of characterizing regional pollutant distributions.

The extent of groundwater contamination from landfills and storage tank leakage is often unknown (Figure 7.18). It is important to devise measurement strategies to characterize the spatial, chemical, and temporal nature of this problem. Chemical engineers have been at the forefront in using advanced mathematical tools and instrumentation to characterize the size and extent of petroleum reservoirs. This technology should be transferred to the groundwater problem and, in particular, to the task of designing cost-effective sampling strategies. Other options include probing potential subsurface sources of

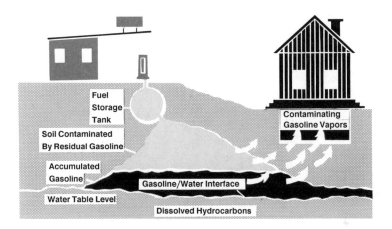

FIGURE 7.18 Leakage from underground storage tanks. Some 2 million underground steel storage tanks are buried beneath service stations throughout the country. Another 1.5 million steel gasoline tanks are buried on farms. Thousands of "orphan" tanks are believed to have been left behind when service stations were razed for redevelopment. Leaks from these tanks could allow hazardous organic compounds to migrate beneath the surface, polluting soil and aquifers and releasing fumes to the surface. Courtesy, E.I. du Pont de Nemours and Company, Inc.

pollutants by nondestructive methods such as acoustic probing, eddy current techniques to assess tank corrosion, magnetometers for location of buried drums, and electrical resistivity measurements.

Improvements in the monitoring and operation of incinerators could minimize the accidental release of hazardous effluents. In particular, fast-acting, continuous, on-line monitors are needed to detect excursions in operating conditions that could lead to toxic emissions.

Development of methodologies for characterizing and measuring human exposure to chemicals is a challenging scientific and engineering undertaking. Data are needed for studies of risk assessment and health effects. During the past decade, rudimentary monitors have become available to determine a person's exposure by measuring concentrations of a given pollutant in the air breathed. Efforts should be directed to lowering the cost and increasing the sensitivity, chemical selectivity, and accuracy of these monitors. Widespread use of personal exposure monitors offers the potential for improving epidemiological studies and for developing a more rigorous scientific basis for setting standards in the workplace and the general environment.

Multimedia Approach to Integrated Chemical Management

Current laws and programs focus on the removal of pollutants from the medium—air, water, or land—in which they are found, often with little regard for chemical management of the environment as a whole. Because modern analytical techniques have revealed trace amounts of many toxic chemicals throughout the environment, however, the medium-specific approach to pollution control is now questionable.[15] The diverse effects of acid rain and of leachates from hazardous waste sites illustrate the mobility of chemicals in the environment (Figure 7.19). The following list gives many areas of research opportunity in a multimedia approach to chemical management of the environment:

- characterization of background levels of chemicals and geochemical cycles;
- basic kinetic studies of chemical degradation;
- field experiments to measure pollutant fluxes among media;
- fundamental studies of interfacial dynamics;
- better determination of Henry's Law constants for volatile species;
- study of chemical speciation, including binding of organic compounds in soil and surface waters (Figure 7.20);
- studies of sorption and desorption of heavy metals on suspended organic matter; and
- formation of precipitates in response to changes in pH.

ASSESSMENT AND MANAGEMENT OF HEALTH, SAFETY, AND ENVIRONMENTAL RISKS

Two of the biggest challenges facing chemical engineering in the near future are (1) the iden-

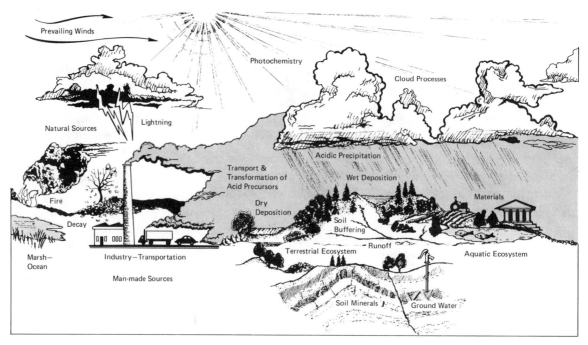

FIGURE 7.19 Mobility of acid rain—a multimedia problem. Reprinted from *Opportunities in Chemistry*, National Academy Press, 1985.

tification and evaluation of the risks—both real and perceived—to human health and to the environment from exposure to chemicals (risk assessment) and (2) the adequate control of these risks (risk management). The ultimate objective in meeting these challenges is to ensure that risks in the chemical and processing industries are viewed as acceptable by the public, regulatory bodies, and the courts, while maintaining worldwide technological leadership and cost competitiveness in these industries by capitalizing fully on advances in chemistry, biotechnology, materials, and microelectronics.

These challenges are critical to the profession of chemical engineering, the chemical industry, and our country. Risk assessment and management involve input from a multitude of different disciplines. The methodology is rapidly changing and extremely complex and requires both technical input and input from professionals with expertise in legal, economic, judicial, medical, regulatory, and public perception issues.

Risk Assessment

Risk assessment, an obvious precursor to risk management, first identifies a hazard and then quantifies the likelihood of occurrence (hazard assessment) and the impact (exposure assessment) associated with each hazard event.

Hazard Identification and Assessment

Two main hazards associated with chemicals are toxicity and flammability. Toxicity measurements in model species and their interpretation are largely the province of life scientists. Chemical engineers can provide assistance in helping life scientists extrapolate their results in the assessment of chemical hazards. Chemical engineers have the theoretical tools to make important contributions to modeling the transport and transformation of chemical species in the body—from the entry of species into the body to their action at the ultimate site where they exert their toxic effect. Chemical engineers are also more likely than life scientists to appreciate realistic conditions and exposure scenarios for the use of hazardous chemicals in industrial settings. Their assistance in interdisciplinary efforts is needed to relate toxicity measurements to actual practice.

Identification of hazards of unexpected, episodic events such as transportation accidents, equipment failure, fires, and explosions is largely the responsibility of the chemical engineer. The most difficult problem in the quantification of toxicity, explosion, and fire hazards of unexpected and episodic natures is the estimation of the probability of occurrence. Risk analysts draw on both analysis and experience to generate sequences of component and subsystem failures that might lead to significant accidents. Methods such as fault-tree analysis can be used to display failure sequences in a logical format. Mechanical failures, operator errors, and management system deficiencies must all be considered in identifying and quantifying risks.

Risk assessment for complex systems involving hazardous chemicals requires deep understanding of the full spectrum of operations. It is highly interdisciplinary and potentially requires manipulation of massive amounts of information, some of which may be missing or of uncertain validity. While the techniques and governing rules for risk assessment are generally straightforward, much creative work needs to be done before the methodology can be used efficiently and effectively to anticipate and correct safety problems or to analyze operating abnormalities for precursors of danger.

New techniques employing expert systems for analysis of complex chemical processes can be used to anticipate safety problems associated with various design decisions. In many major accidents, the relevant fundamental phenomena involved were totally unanticipated and were understood only after considerable investigation. Some of these abnormal conditions might have been identified by a priori research guided by techniques for anticipating interactions that might be overlooked in conventional approaches to design. Other serious accidents were pre-

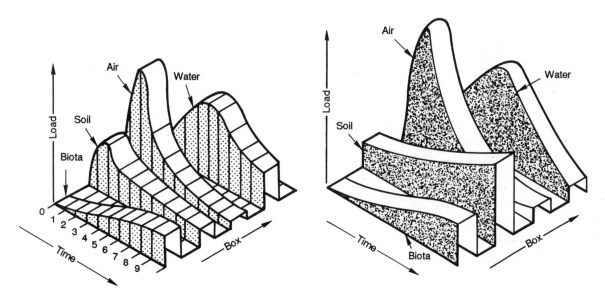

FIGURE 7.20   The binding properties of chemicals can affect their distribution and retention in various environmental media. For example, the hypothetical environmental fates of two chemicals are contrasted. The profile on the left shows the loading in various media over time for a chemical that binds strongly with lipid material. The profile on the right shows the loading in various media over time for a chemical that binds strongly with organic material. Courtesy, Office of Technology Assessment.

ceded by abnormalities in operation that were ignored until it was too late. Again, techniques for identifying and investigating such warning signals might have avoided disaster.

Exposure Assessment

There is a dearth of information and methodology in the area of human and environmental exposure assessment. Standardized methodologies have not been developed; monitoring of personnel exposure is rare; and suitably sensitive, rapid-response, portable analytical equipment is limited. Fortunately, the exposure factor is a controllable variable, at least theoretically, and is largely within the province of the chemical engineer, who selects and designs processes, sites and expands plants, and develops plant operating procedures and transport systems. To determine the likely dispersion of flammable materials, as well as the transport of toxic chemicals in the environment, dispersion/reaction models (both near field and far field) for realistic accident scenarios (e.g., heavy gases, dusts, aerosols) must be developed and verified experimentally.

Risk Management

Once the hazard and exposure assessments are complete for any specific hazard, it is relatively simple to determine how many people will be affected and the severity of the effect (i.e., the risk). It is considerably more difficult to decide whether these risks are warranted compared to the benefits. This is particularly true if the risks are uncertain, involuntary, or not understood by those at risk; if those at risk are not primarily those who benefit; or if alternatives are unknown, uncertain, or impractical. The process is complex because the goals are multiple and frequently contraindicating.

Economic, liability, public image, and opinion considerations are involved. Catastrophic hazards are less acceptable than smaller ones even if the absolute risk is identical. Voluntary risks are a way of life for most people, but there is minimal tolerance for involuntary risks, particularly if they are unknown or not understood.

In today's heavily regulated and litigious society, it is becoming increasingly essential that risk assessments be conducted and adequately and carefully documented for all existing industrial plants and transport systems that handle or store significant quantities of toxic or flammable chemicals. The same must be done in siting new facilities, selecting processes, designing processes and equipment, developing operating and maintenance procedures, and designing transport systems.

Policies and procedures for risk management decisions must be established and be clear and simple if the massive, but necessary, workload of risk assessment and management is not to cripple the chemical industry's worldwide competitive position and consume inordinate resources through inefficiency.

Managing chemical risk must proceed in the absence of perfect information on risks and how to avoid them. The lack of critical information or good alternatives is no excuse for inaction.

IMPLICATIONS OF RESEARCH FRONTIERS

This chapter has made clear the challenges to chemical engineers in research related to environmental protection, process safety, and hazardous waste management. Chemical engineering education must become strongly oriented to these topics, as well. For example, what might be characterized as the traditional approach to environmental concerns in process design—establishing the process and then providing the necessary safety and environmental controls—must give way to a new approach that considers at the earliest stages of design such factors as

- process resilience to changes in inputs,
- minimization of toxic intermediates and products, and
- safe response to upset conditions, startup, and shutdown.

As chemical engineering research develops new design and control tools to deal with these factors, these tools should be integrated into the curriculum. Process safety research is generally more advanced in industry than it is in academia. Closer interaction between industrial researchers and academic researchers and educators is needed to disseminate insights and knowledge gained by industry in this area.

Other problems in environmental science and technology—as well as an introduction to the social, economic, and political aspects of environmental issues—should receive broad exposure in the curriculum. They should be in the content of existing courses wherever possible.

Industry has strong developmental research programs in areas such as process safety, but more fundamental research on process design tools, on emerging environmental problems, or on general topics linked to public health and environmental protection requires stable, long-term research support from the federal government. The chemical engineering profession stands ready to tackle these issues aggressively; does the federal government?

The principal federal agencies that have supported environmental research generally have been the Environmental Protection Agency, the National Science Foundation, the National Oceanic and Atmospheric Administration, the National Institute of Environmental Health Sciences, and the Department of Energy. In recent years, many of these agencies have experienced budget cuts that threaten their ability to maintain vital research programs that anticipate environmental problems, instead of reacting to the latest crisis. Cutting back on basic research on environmental problems is a false economy. Small savings today on anticipatory research may result in very large costs to society in the future, since dealing with the consequences of environmental problems is invariably more expensive than research to anticipate and prevent these problems.

Because of the critical importance of maintaining our environmental quality and improving process safety and hazardous waste management, the committee recommends that these federal agencies undertake new initiatives in chemical engineering research. The details of proposed initiatives for EPA and NSF are spelled out in more detail in Chapter 10 and Appendix A of this report. An investment in research to anticipate and prevent environmental problems is likely to be highly cost-effective. The costs

of responding to unforeseen environmental problems have certainly been great. Significantly increased support for fundamental research is vital if universities are to preserve their role in long-term environmental research and in the education of tomorrow's researchers, process designers, and regulators.

NOTES

1. J. L. Makris. *Natural Hazards Observer*, 10(3), January 1986, 1.
2. J. McLoughlin. "Risk and Legal Liability" in *Dealing with Risk*, R. F. Griffiths, ed. New York: John Wiley & Sons, 1982.
3. National Safety Council. *Accident Facts*. Chicago: National Safety Council, 1985.
4. U.S. Department of Commerce, Bureau of the Census. *Statistical Abstract of the United States: 1987*, 107th ed. Washington, D.C.: U.S. Government Printing Office, 1986, Table 697.
5. *One Hundred Largest Losses—A Thirty-Year Review of Property Damage Losses in the Hydrocarbon-Chemical Industries* (OHL-9–86–7). Chicago: Marsh and McLennan, 1986.
6. McGraw-Hill Economics. *Survey of Investment in Employee Safety and Health*, 13th ed. New York: McGraw-Hill, 1985.
7. James L. Regens, "The Regulatory Environment for Coal Development," in *Costs of Coal Pollution Abatement: Results of an International Symposium*. Paris: Organisation for Economic Cooperation and Development, 1983.
8. *The National Survey of Hazardous Waste Generators and Treatment, Storage, and Disposal Facilities Regulated Under RCRA in 1981*. WESTAC, Inc., 1984.
9. U.S. Congress, Congressional Budget Office. *Hazardous Waste Management: Recent Changes and Policy Alternatives*. Washington, D.C.: Congressional Budget Office, May 1985.
10. U.S. Congress, Committee on Public Works and Transportation, Subcommittee on Investigations and Oversight. *Hazardous Waste Contamination of Water Resources (Concerning Groundwater Contamination in Santa Clara Valley, CA)* (99-32). Washington, D.C.: U.S. Government Printing Office.
11. U.S. Congress, Office of Technology Assessment. *Superfund Strategy* (OTA-ITE-252). Washington, D.C.: U.S. Government Printing Office, 1985.

12. T. A. Kletz. *Simpler, Cheaper Plants or Wealth and Safety at Work—Notes on Inherently Safer and Simpler Plants*. London: Institution of Chemical Engineers, 1984.

13. U.S. Environmental Protection Agency, Science Advisory Board. *Incineration of Hazardous Liquid Waste*. Washington, D.C.: U.S. Environmental Protection Agency, 1984.

14. U.S. Congress, Office of Technology Assessment. *Technologies for Hazardous Waste Management*. Washington, D.C.: U.S. Government Printing Office, 1981.

15. For a detailed discussion of this topic, see Y. Cohen, ed. *Pollutants in a Multimedia Environment*. New York: Plenum Press, 1986.

EIGHT

Computer-Assisted Process and Control Engineering

Computers and computational methods have advanced to the point where they are beginning to have significant impact on the way in which chemical engineers can approach problems in design, control, and operations. The computer's ability to handle more complex mathematics and to permit the exhaustive solution of detailed models will allow chemical engineers to model process physics and chemistry from the molecular scale to the plant scale, to construct models that incorporate all relevant phenomena of a process, and to design, control, and optimize more on the basis of computed theoretical predictions and less on empiricism. This chapter explores the implications of these changes for process design, control, operations, and engineering information management. Future developments and opportunities in process sensors are also covered.

IMAGINE A ROOM 50 feet long and 30 feet wide, filled with cabinets containing 18,000 vacuum tubes interconnected by miles of copper wire. This was the ENIAC (Figure 8.1), a 30-ton behemoth that was the world's first electronic computer. Now imagine that ENIAC had been given the task of solving a set of simultaneous linear equations that embodied 30,000 independent variables. With a team of people working around the clock to record intermediate solutions and feed them back into the computer (ENIAC had a limited amount of memory), ENIAC would still be chugging away more than 40 years later to solve the problem! A supercomputer using modern algorithms could solve the problem in an hour or so.

The speed and capability of the modern computer have tremendous implications for the practice of chemical engineering. In the future, computer programs incorporating artificial intelligence or expert systems will help engineers design improved chemical processes more efficiently. Complex computations based on fundamental engineering knowledge will allow engineers to design reactors that can virtually eliminate undesired by-products, making processes less complex and less polluting. New sensors, many of which will be miniature analytical laboratories tied to miniature electronics, will allow rapid and accurate measurements for control that are currently impossible. New chemical products that today are discovered predominantly through laboratory work—for instance, reinforced plastics that are as strong as steel and weigh less than aluminum or drugs with miraculous properties—may be discovered in the future by computer calculations based on models that predict the detailed behavior of molecules.

Chemical engineers will lead this revolution. They will need to be trained to use advanced computer techniques for process design, process control, and management of process information. Advanced engineering development will be based more than ever on mathematical modeling and scientific computation. Reliable modeling at the microscale, the individual process unit scale, and the plant process scale will improve our ability to scale up processes in a few large steps, possibly bypassing the need for a pilot plant and saving the 2 or 3 years required to build and operate it. Process models capable of predicting dynamic behavior, operability, flexibility, and potential safety problems will permit these aspects of a process to be considered more fully earlier in the design stage. Because improved computers can perform the extensive computations required by such models, it will be possible to test alternative designs more quickly.

A chemical process must be designed to operate under a chosen set of conditions, each of which must be controlled within specified limits if the process is to operate reliably and yield a product of specified quality. Accurate, complex, computer-solvable models of chemical processes will incorporate features of the controls that are needed to maintain the desired process conditions. Such models will be able to

FIGURE 8.1 The Electronic Numerical Integrator and Calculator (ENIAC) and its inventor, J. Presper Eckert, circa 1946. ENIAC was the world's first electronic computer. Courtesy, UNISYS Corporation.

predict the effects of process excursions and the control measures needed to correct them. Computer management of the process operation will rapidly initiate control correction of process excursions. The development of new types of process sensors will be essential to this degree of process control and will eliminate the time-consuming withdrawal of process stream samples for analysis.

The design of a commercial process can generate an almost uncountable number of possible solutions for seemingly simple problems. Even a decade's 10,000-fold increase in computational capability in terms of faster computer speeds and better algorithms does not permit a person to search among these alternatives, nor would such a search make strategic sense. Give engineers a design problem for which the best solution is obvious from today's technology, and they will quickly write down the correct solution without searching. If the solution is not obvious, they will often home in on the information needed and perform the computations that will expose the right solution quickly and with minimal effort. By using intuition and experience, they eliminate the need for testing every possible alternative. We need to understand how to use computer technology in much the same way, to solve complex problems where many of the decisions are based on qualitative information and insights that develop as the problem is attacked.

Encoding this activity in the computer involves a type of modeling in which the capabilities of the designer and his tools, the alternative procedures by which complex problem solving can be performed, and effective methods of information management are all incorporated. Advances in artificial intelligence, expert systems, and information management will revolutionize the automation of this activity, giving us computers that can display encyclopedic recall of relevant information and nearly human reasoning capabilities. A HAL 9000 of *2001: A Space Odyssey* fame may indeed exist in the future.

Computer-generated visual information, for example, three-dimensional portrayal of proposed new designs, will be commonplace in the future. Communication will be in natural language, using both pictures and voice. This setting, which will address the need for new chemical processes and products by harnessing almost unimaginable computing power, will provide significant new research opportunities in chemical engineering.

USING THE COMPUTER'S POTENTIAL

Each decade over the last 35 years has seen the processing speed of newly designed computers increase by a factor of about 100 owing to advances in the design of electronic microcircuits and other computer hardware. On top of this has been another 100-fold increase per decade in computer speed, thanks to more efficient methods of carrying out computations (algorithms). It is not widely appreciated that new algorithms have been as valuable as hardware design in improving computer performance. With the combination of improved computer hardware and better algorithms, effective computer speeds have more than doubled on average each year.

The availability of computing resources is also increasing rapidly. The actual and projected availability of high-speed supercomputers is shown in Table 8.1. The projection for 1990—at least 700 computers of Cray I class—represents 35 times more available computing power than that available in 1980. While continued substantial investment in supercomputers is needed, support for better ways of using them should not be neglected. It is conceivable that a new algorithm could effectively increase the power of supercomputing for a specific problem by a factor of 35 overnight. During the last two decades, many developments in numerical anal-

TABLE 8.1 Actual and Estimated Supercomputing Resources Available to Researchers in the United States, 1980–1990

Year	Number of Cray I-Class Supercomputers
1980	21
1985	142
1990	700–1,000

ysis have had a profound impact on scientific computation.

Clearly computer technology has improved rapidly; there is little reason to doubt that it will continue to do so. The problem is now and will continue to be the lack of people trained to apply computer technology to scientific and engineering tasks. The improvements suggested in Chapter 7 in terms of our ability to design and control better chemical products and processes will be made by chemical engineers who understand computers—not by computer scientists or by software engineers. The countries that understand this distinction will lead the world in chemical technology.

MATHEMATICAL MODELS OF FUNDAMENTAL PHENOMENA

Chemical engineers have traditionally used mathematical models to characterize the physical and chemical interactions occurring in chemical processes. Many of these models either have been entirely empirical or have relied on crude approximations of the basic physics or chemistry of the process. This is because a typical chemical process comprises an assemblage of interacting flows, transports, and chemical reactions. Accurate analysis and prediction of the behavior of such a complex system require detailed portrayal of the physics of transport and the chemistry of reactions, which calls for complex equations that do not yield to traditional mathematics. Nonlinear partial differential and integral equations in two and sometimes three spatial variables must be solved for regions with complicated shapes that often have at least some free boundary. The more accurate the model, the more mathematically complex it becomes, but it cannot be more complex than allowed for

Process Design in the Twenty-First Century

It is April 2008. A process engineer for a medium-size chemical company is designing a process for a new series of products her company is developing. By talking to her computer and pointing to the screen, she indicates that the process needs a reactor and some separators, as well as recycle capability. The computer, using an artificial intelligence program tied to mathematical models, determines how the reactor temperature and pressure will influence the product yield and how the yield in turn will influence the amount of process stream recycle. The computer chooses the optimal types of separators and indicates how they should be sequenced for different process streams and products of the series. After 20 minutes on a teraflop machine (equivalent to 1 week of computing on a CRAY I of the 1980s), the computer indicates that the general process design is completed, having given partial information during the 20-minute computation.

The graphical output from the computer shows the process flowsheet, with several separation units and projected equipment and operating costs. It also flags information that is uncertain because it had to use thermodynamic data extrapolated from measured values. At the engineer's request, the computer shows several alternative flowsheets it had considered, indicates their projected costs, and tells why it eliminated each of them. Some of the flowsheets were eliminated because of high cost, others because they were considered unsafe, others because the startup procedures would be difficult, and still others because they were based on uncertain extrapolation of experimental data.

After perusal of these process options, the engineer asks the computer to select five designs for further study, and the computer produces a paper copy of the flowsheet and design parameters

by the available methods for solving its equations.

Before the advent of modern computer-aided mathematics, most mathematical models of real chemical processes were so idealized that they had severely limited utility—being reduced to one dimension and a few variables, or linearized, or limited to simplified variability of parameters. The increased availability of supercomputers along with progress in computational mathematics and numerical functional analysis is revolutionizing the way in which chemical engineers approach the theory and engineering of chemical processes. The means are at hand to model process physics and chemistry from the for each. She then tells the computer to prepare more rigorous designs from each of the five flowsheets. The computer questions the third design because it involves extensive extrapolation of known thermodynamic data. However, since the engineer feels that this may be one of the better designs, she asks the computer to notify the laboratory management computer, which is connected to a laboratory technician's terminal, to begin experimental determination of the missing thermodynamic data so that the design can be prepared. The computer reminds the engineer that precise control of process conditions will be essential for another of the designs and that a new sensor may be necessary. The engineer consults the artificial intelligence program, which tells her that either of two sensors being developed in another company department may fill her need. The engineer uses her computer to schedule a meeting with the manager of sensor development.

Two years later, the process is producing one of the new products. The time from design to operation was short enough that the company was able to capture a new market, and it plans to extend the business with related products of the series. The design engineer has developed a real-time dynamic simulation of the process and is running it on the company's parallel supercomputer to test strategies for producing the related products from the process equipment. She queries another computer to search the historical records of plant operation. She wants to know how well the dynamic model mirrors the actual plant operation and when the process parameters were last adjusted to give a better fit between the plant and the model. She feeds this information to her computer, which devises a comprehensive plan for scheduling production among four of the new products, with provision for revising the schedule as inventories and sale of the products change. A new business has been created.

molecular scale to the plant scale; to construct models that incorporate all relevant phenomena of a process; and to design, control, and optimize more on the basis of computed theoretical predictions and less on empiricism. Chemical engineers, using advanced computational methods and supercomputers, can now readily identify the important phenomena in a complex chemical process over the entire range of applicable conditions by exhaustive solution of detailed models. The benefits of investing in less empirical, more fundamental mathematical models are becoming clear:

• The capability to construct mathematical

models that more fully incorporate the basic chemistry and physics of a system provides a mechanism for assessing understanding of fundamental phenomena in a system by comparing predictions made by the model with experimental data.

• Better models can replace laboratory or field tests that are difficult or costly to perform or identify crucial experiments that should be carried out. In either case, they will significantly enhance the scope and productivity of chemical engineering researchers in academia and industry.

• In process design, it is frequently discovered that many of the basic data needed to understand a process are lacking. Because most current mathematical models are not sufficiently accurate to permit direct scale-up of the process from laboratory data to full plant size, a pilot plant must be constructed. As models are improved, it may become possible to evaluate design decisions with more confidence, and bypass the pilot plant stage.

Process technologies for which the use of more comprehensive mathematical models can result in major improvements include those for biochemical reaction processing; high-performance polymers, plastics, composites, and ceramics; chemical reaction processing (e.g., reaction injection molding, reaction coating, chemical vapor deposition); microelectronic circuits; optical fibers and disks; magnetic memory systems; high-speed coating; photovoltaic and semiconductor materials; coal gasification; enhanced petroleum recovery; solution mining; and hazardous waste disposal. To date the most extensive use of supercomputer modeling has been in space age weapons technologies, where objectives, economics, and time frames differ from those in the chemical process industries. It is clearly in the national interest to stimulate the more extensive use of advanced computational methods and supercomputers in other industries critical to our worldwide competitive position. A program to encourage the greater dissemination of advanced computational techniques and hardware will offer challenges and opportunities to computational mathematicians and numerical analysts, to engineering scientists, to applications and software experts in firms that develop and manufacture supercomputers, and, above all, to perceptive leaders in high-technology process industries.

The following sections describe in more detail a number of areas in chemical engineering in which the ability to develop and apply detailed mathematical models should yield substantial rewards.

Hydrodynamic Systems

Much of the current computational modeling research in chemical engineering is concerned with the behavior of flowing fluids. The general system of equations that describe fluid mechanics, called the Navier-Stokes equations, has been known for more than 100 years, but for complex phenomena the equations are exceedingly difficult to solve. Only recently have methods been devised to treat such phenomena as shock waves and turbulence. Further difficulties arise when disparate temporal and spatial scales are present and when chemical reactions occur in the fluid. Solutions of the Navier-Stokes equations can be smooth and steady, or they can exhibit regular oscillations or even chaos. In some cases the fluid flow is enclosed by a rigid boundary with a complex shape, as in the extrusion of polymers; in others the flow is effectively un-

Hypercubes

One thing that has not changed since the earliest days of the computer age is this: the fastest available computer is never fast enough to solve the most difficult problem of the day. Because of the limitations of current computers, simplifying assumptions about most physical systems must be made for the problem to be solved by available computer technology. For example, in petroleum reservoir simulation, available computer power dictates how many grid points can be used to describe the reservoir and how complex the thermodynamic model that describes the system can be. A computer capable of solving these complex systems without resorting to simplifying assumptions would have to be orders of magnitude faster than those available today.

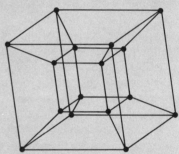

Computer manufacturers have traditionally boosted performance by using faster components. But the overall architecture of computers has not changed since the 1940s, and the use of faster components is a strategy that is now running into diminishing returns. New computer architectures, based on parallel processing of subelements of the same problem, will be needed to achieve significant advances in speed to solve the complex problems of today and tomorrow.

bounded and the solution must extend to infinity, as in atmospheric systems; and in still others, such as the flow of blood in vessels, the boundary is deformable. Solution of the Navier-Stokes equations for systems of technological interest remains an exceedingly challenging task; supercomputers are needed to treat those systems that can be solved.

Polymer Processing

The development of polymers and polymer composites will benefit greatly from the availability of better computers and better algorithms. The inherent properties of a polymer are governed by the chemical structure of its molecules, but the properties of a finished polymer product are affected by the interactions

One example of such a new approach is the hypercube architecture, in which many processors are linked together as a team to solve a single problem. A well-integrated team of cheap processing units can potentially outperform the most sophisticated single-processor machine. In addition, since each processor can have its own dedicated memory, the total system can have both more memory than current supercomputers and more memory in use at any given moment.

Standard programs must be broken into smaller pieces to run on a hypercube. Each processor is assigned the responsibility for calculations for a specific piece of a problem. For example, in petroleum reservoir simulation, each processor might be assigned a different section of the reservoir to model. In modeling a complex chemical plant, each processor might be assigned a different piece of equipment. As each processor proceeds, it informs the other processors of its results, so that all the other processors can incorporate the information into their respective portions of the overall calculation.

Hypercubes and other new computer architectures (e.g., systems based on simulations of neural networks) represent exciting new tools for chemical engineers. A wide variety of applications central to the concerns of chemical engineers (e.g., fluid dynamics and heat flow) have already been converted to run on these architectures. The new computer designs promise to move the field of chemical engineering substantially away from its dependence on simplified models toward computer simulations and calculations that more closely represent the incredible complexity of the real world.

physicists, petroleum engineers, and chemical engineers. As more sophisticated techniques are developed for locating and recovering petroleum, mathematical modeling is playing an ever-expanding role.

Once regions that may contain petroleum are located, local geological features must be sought that might have trapped the hydrocarbons. The discovery process is based on a kind of seismic prospecting in which geologic maps are constructed from reflected seismic signals generated by explosions or vibrations at the surface of the earth. These signals are reflected or refracted in varying degrees by different rock strata and are recorded by a set of receivers. Thus, the problem of interpreting signals can be likened to that of analyzing light beams reflected by an array of variously curved plates of glass of different reflectivities separated by liquids of different refractive indexes. The inverse mathematical problem of determining the earth structure and the properties of the strata from the recorded signals is extremely difficult.

among these molecules, which are strongly influenced by the way in which the material has been processed. While it is now possible to predict certain properties of polymers from their molecular structure, the ability to predict the effect of polymer processing steps on polymer properties is just being developed. Ideally it would be desirable to model all steps from the formation of the polymer through its processing and then predict the final properties of the material from structure-property relationships. Although such modeling is a formidable problem, it is becoming feasible with the advent of supercomputers and improved algorithms.

Petroleum Production

Computation is widely used in petroleum exploration and production by exploration geo-

After a hydrocarbon reservoir has been located, the flow of oil, water, gas, and possibly injected chemicals in the reservoir must be modeled. This challenge is particularly appropriate for chemical engineers working with petroleum engineers because of the important role played by molecular level interactions between oil, subsurface water, and rock. Models for fluid flow in porous media comprise coupled systems of nonlinear partial differential equations for conservation of mass and energy, equations of state, and other constraining relationships. These models are usually defined on irregular domains with complex boundary conditions. Their numerical solution, with attendant difficulties such as choice of discretization methods and grid orientation, is a challenging intellectual problem.

Once wells have been drilled into the formation, the local properties of the reservoir rocks and fluids can be determined. To construct a realistic model of the reservoir, its properties over its total extent—not just at the well sites—must be known. One way of estimating these properties is to match production histories at the wells with those predicted by the reservoir model. This is a classic ill-posed inverse problem that is very difficult to solve.

When or if the reservoir is successfully simulated, the engineer can turn to optimizing petroleum recovery, and theoretical ideas can be applied to models for various enhanced recovery methods to select optimal procedures and schedules (see Chapter 7).

Combustion Systems

Combustion is one of the oldest and most basic chemical processes (Plate 6). Its accurate mathematical modeling can help avoid explosions and catastrophic fires, promote more efficient fuel use, minimize pollutant formation, and design systems for the incineration of toxic materials (see Chapter 8). For example, modeling the initiation and propagation of fires, explosions, and detonations requires the ability to model combustion phenomena. Models of the internal combustion engine can shed light on the influence of combustion chamber shape or valve and spark plug placement on engine performance. Models at the molecular level can provide a fundamental understanding of how fuels are burned and how gaseous and particulate pollutants are formed. This can lead to ways to improve the design of combustion systems.

Mathematical models of combustion must incorporate intricate fluid mechanics coupled with the kinetics of many chemical reactions among a multitude of compounds and free radicals. They must also consider that those reactions are taking place in turbulent flows inside chambers with complex shapes. Because complete models of real combustors, incorporating accurate treatment of both fluid mechanics and chemistry, are still too large for present computers, the challenge is to construct simpler, yet still valid, models by using critical insight

into the important chemical and physical phenomena found in combustors. Chemical engineers have the mix of expertise necessary to accomplish this.

Environmental Systems

The environment can be likened to a giant chemical reactor. Gases and particles are emitted into the atmosphere by industrial and other man-made processes, as well as by a variety of natural processes such as photosynthesis, vulcanism, wildfires, and decay processes. These gases and particles can undergo chemical reactions, and they or their reaction products can be transported by the wind, mixed by atmospheric turbulence, and absorbed by water droplets. Ultimately, they either remain in the atmosphere indefinitely or reach the earth's surface. For example, the hazes of polluted atmospheres consist of submicron aerosols of inorganic and organic compounds, which are formed by chemical reaction, homogeneous nucleation, or condensation of gases (Plate 7).

Models of atmospheric phenomena are similar to those of combustion and involve the coupling of exceedingly complex chemistry and physics with three-dimensional hydrodynamics. The distribution and transport of chemicals introduced into groundwater also involve a coupling of chemical reactions and transports through porous solid media. The development of groundwater models is critical to understanding the effects of land disposal of toxic waste (see Chapter 7).

PROCESS DESIGN

The primary goal of process design is to identify the optimal equipment units, the optimal connections between them, and the optimal conditions for operating them to deliver desired product yields at the lowest cost, using safe process paths, and with minimal adverse impact to the environment. Design is a complex problem that involves not only the quantitative computing depicted in the previous section, but also the effective handling of massive amounts of information and qualitative reasoning.

Computer-Assisted Design of New Processes

Designs for new processes proceed through at least three stages:

- *Conceptual design*—the generation of ideas for new processes (process synthesis) and their translation into an initial design. This stage includes preliminary cost estimates to assess the potential profitability of the process, as well as analyses of process safety and environmental considerations.
- *Final design*—a rigorous set of design calculations to specify all the significant details of a process.
- *Detailed design*—preparation of engineering drawings and equipment lists needed for construction.

The key step in the conceptual design of a new chemical manufacturing process is generating the process flowsheet (Figure 8.2). All other elements of computer-aided design (e.g., process simulation, design of control systems, and plantwide integration of processes) come into play after the flowsheet has been established. In current practice, the pressure to enter the market quickly often allows for the exploration of only a few of the process alternatives that should be considered. To be fair to today's designers, it is possible to generate a very large number of alternative process paths at the conceptual stage of design, and yet experience indicates that less than 1 percent of the ideas for new designs become commercial. Thus, the challenge in computer-aided process synthesis is to develop systematic procedures for the generation and quick screening of many process alternatives. The goal is to simplify the synthesis/analysis activity in conceptual design and give the designer confidence that the initial universe of potential process paths contained all the pathways with reasonable chances for commercial success. The advances in computer-aided process synthesis that are possible over the next decade are dramatic. They include both an increasing level of sophistication (e.g., the synthesis of heat exchanger networks, sequences of separation processes, networks of reactors, and process control systems) and computational procedures that should make possible

the identification of the most viable process option in a relatively short amount of time.

As the designer moves from conceptual design toward final design, he or she must analyze a number of alternatives for the final design. The development of large, computer-aided design programs (so-called process simulators) such as FLOWTRAN, PROCESS, DESIGN 2000, and ASPEN (or other equivalent programs used in various companies) has significantly automated the detailed computations needed to analyze these various process designs. The availability of process simulators has probably been the most important development in the design of petrochemical plants in the past 20 years, cutting design times drastically and resulting in better designed plants.

Although the available simulators have done much to achieve superior design of petrochemical processes, there is considerable room for improvement. For example, better models are needed for complex reactors and for solids processing operations such as crystallization, filtration, and drying. Thermodynamic models are needed for polar compounds. Moreover, the current process simulators are limited to steady-state operations and are capable of analyzing only isolated parts of a chemical plant at any given time. This compartmentalization is due to the limitations on computer memory that prevailed when these programs were first developed. This memory limitation resulted in a computational strategy that divided the plant into "boxes" and simulated static conditions within each box, iteratively merging the results to simulate the entire plant. With today's supercomputers, it is possible to simulate the dynamics of the entire chemical plant. This opens the way for dramatic advances in modeling and analysis of alternative process designs, because the chemical reactions that occur in manufacturing processes are usually nonlinear and interdependent, and random disturbances in the process can propagate quickly and threaten the operation of the entire plant. To nullify the effects of such disturbances, the designer must know the dynamics of the entire plant, so that control failure in any one unit does not radiate quickly to other units. It is now within our reach to integrate this sophisticated level of

144

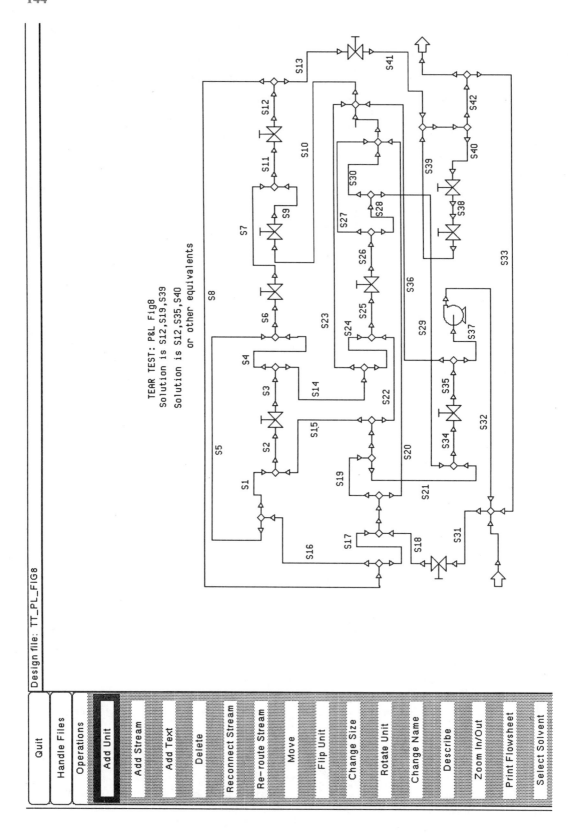

FIGURE 8.2 Example of a flowsheet generated by computer-aided process synthesis, as would be seen on the screen of a computer terminal. The column on the left side of the figure shows options available in the particular design program being run. Courtesy, Peter Piela, Carnegie-Mellon University.

design and analysis on a plantwide scale (including design and performance modeling of plantwide control systems) into the computational tools used to analyze and optimize the performance of individual processes in the plant.

In the detailed design stage for a chemical manufacturing process, a chosen process design must be converted into a list of equipment items to be purchased and a set of blueprints to guide their assembly. The design is presented as a detailed process flow diagram (PFD), from which is constructed a list of all needed items of equipment, and piping and instrumentation diagrams (PIDs) that show the equipment and its interconnections. The next task is to establish the physical layout for the entire plant. Advanced computer-based drafting tools aid in all these activities.

Computer-Assisted Process Retrofitting

The preceding section focused largely on the design of new plants. However, these procedures can also be adapted to the retrofitting of existing plants. Retrofitting is generally undertaken to increase the capacity of a plant; to make use of a new technology such as an improved catalyst, a new material of construction, or a new unit operation; or to respond to a significant change in the cost of energy or raw materials. A fair amount of retrofitting in the chemical process industries in recent years has been undertaken to improve energy efficiency through plantwide energy integration—retrofitting of about 50 processes to incorporate modern heat-exchanger network synthesis concepts has reduced energy requirements in the chemical industry by 30 to 50 percent. Retrofitting will continue to play a major role in the design of chemical plants as new procedures for computer-aided synthesis of separation systems are applied and research on process synthesis begins to yield large savings by helping existing petrochemical plants produce the same mix of products through more economical chemical reaction pathways.

We need to develop a systematic approach to analyzing the impact of making changes in the connections between process units or in the

size of units that are undertaken to improve operating costs, plant flexibility, or safety.

Research Opportunities in Process Design

The overall goal of process design research is to develop a systematic procedure, probably in the form of an interactive computer program, that contains design heuristics and interchangeable approximate and rigorous models that can lead an engineer from an initial concept to a final design as quickly as possible. The final design must include considerations of economics, controllability, safety, and environmental protection. We need to extend the conceptual and final design procedures that have been developed for petrochemical processes to processes for producing polymers, biochemicals, and electronic devices. We also must develop systematic synthesis/analysis procedures for studying batch processes analogous to the procedures that have been developed for studying continuous petrochemical processes.

There are some aspects of process design in which decisions are based primarily on past experience rather than on quantitative performance models. Problems of this type include the selection of construction materials, the selection of appropriate models for evaluating the physical properties of homogeneous and heterogeneous mixtures of components, and the selection of safety systems. Advances in expert systems technology and information management will have a profound impact on expressing the solutions to these problems.

In summary, systematic procedures must be developed for the following:

• generation of process alternatives;

• quick screening of process alternatives using both rule-of-thumb and short-cut calculations;

• inclusion of controllability, safety, and environmental factors in the initial design;

• more detailed screening of a small number of promising design alternatives;

• design of the process and management of its construction;

• use and extension of expert systems concepts to handle aspects of design that deal with a mixture of qualitative and quantitative information; and

- retrofitting of existing plants.

PROCESS OPERATIONS AND CONTROL

Process operations and control have a tremendous impact on the profitability of a manufacturing operation. In some cases, they can determine the economic viability of a manufacturing facility. For example, Du Pont's Process Control Technology Panel has estimated that if Du Pont were to extend the degree of computer process control that has been achieved at a few of its plants across the entire corporation, it would save as much as half a billion dollars a year in manufacturing costs. If Du Pont's numbers are representative, the entire chemical industry could save billions of dollars each year through more widespread application of the best available process control. This could be the single most effective step that the U.S. chemical process industry could take to improve its global competitive position in manufacturing.

Why are such savings suddenly possible? Because of the explosive developments in computer technology, research on operations and control is no longer constrained by lack of computing power. In particular, the traditional boundaries between design, control, optimization, simulation, and operation are disappearing. Control is becoming a part of process design; simulation and optimization are becoming components of control design.

Research opportunities in process operations and control lie in three areas:

- collection of information through process measurements;
- interpretation and communication of information by use of process models; and
- utilization of information through control algorithms and control strategies for both normal and abnormal operation.

Measurements

The essence of process control is to take appropriate and quick corrective action based on measured information about the behavior of the process. The concept of the process is contained in a process model, and measurements are used to evaluate the degree to which the process conditions deviate from those of the model. When a mismatch occurs between actual process conditions and those postulated by the model, a control and operating strategy is invoked to correct the process conditions. A critical interrelationship exists between measurements and operating/control strategy, one that is too often neglected. The control strategy depends on what information is or can be available, even while it dictates which measurements are needed. This is perhaps best illustrated in a number of manufacturing processes in cutting-edge technologies, where control and operating strategy is circumscribed by the lack of appropriate sensors for many critical process variables. Conventional estimation techniques are used that infer the values of unmeasured variables from measured variables, but these provide imperfect guidance for process control.

Even something as empirical as process measurement cannot be divorced from the need for good process modeling. In the absence of a good model, it may not be known what variables affect process operation or product quality and should therefore be measured or estimated.

Process measurements are subject to errors. Random (stochastic) disturbances are ubiquitous, and gross or systematic errors can be caused by malfunctioning sensors or instruments. The detection and elimination of these errors are essential if the data are to be used for process operations and control. The success of this screening depends on the measurements themselves, the failure data available, and the process control strategy. At the present time, diagnostic programs are not applied to most sensor failure data. The detection and remediation of significant errors in measurements for process control pose interesting research opportunities.

Interpretation of Process Information

The quality of an operation and control strategy depends on the quality of the model on which it is based (Figure 8.3). We are only beginning to understand this relationship quan-

"...and in 1/10,000 of a second it can compound the process model's error 87,500 times!"

FIGURE 8.3 The importance of accurate process models. Copyright 1988 by Sidney L. Harris.

titatively, even in the relatively simple context of the feedback control of linear systems. Even if it is assumed that the structure of the process model is correct, we do not yet know how to translate uncertainties in model parameters into uncertainty in the performance of the control system. A more difficult problem is to assess the effect of an incorrect model structure, such as a wrong set of basic equations, on the performance of the control strategy on which it is based. Understanding the effect of model-process mismatch on control system performance provides a critical research opportunity.

In the context of operations and control, simulations can be used to test new process strategies as well as to train operating personnel to control the process and to respond to emergencies. The increasing use of simulation in process control requires that the cost of dynamic simulation be brought down. This could be done by taking advantage of new developments in computers, such as new user interfaces, computer architectures, and languages, and by developing faster numerical integration algorithms for ordinary differential equations.

Alarm management also requires research. Modern chemical plants usually have audible and visual annunciators to warn operators when key variables deviate from acceptable or safe values. A process upset in a plant that has several interconnected units with many feedback controls can set off multiple alarms, and the consequences of misinterpreting the alarms can range from inefficient process operation to outright disaster. When the alarm sounds, the operator must decide quickly what action to take. A hybridization of expert systems and process control systems can assist the operator in interpreting process status after an abnormal event. The need for better handling of abnormal events makes research in artificial intelligence of great importance to the chemical industry.

Integration of Process Design with Control

Most continuous plants are now designed for steady-state operation with little regard for the ease (or difficulty) with which the steady state can be maintained through control. Such a plant can be difficult to control once it deviates from the steady state.

Design and control have traditionally been treated separately for the following reasons:

- The problems in each area alone are exceedingly complex.
- The interactions between design, control, and optimization are poorly understood.
- The computational requirements of an integrated approach to design and control have been beyond the capability of available hardware and software.

For example, a chemical plant might be designed to achieve high efficiency by integrating the operation of many individual process units across the plant (e.g., by using waste heat from one unit as an energy source in another unit). However, the tight coupling of process units generally makes the entire plant more difficult to control. Therefore, this is a factor that must be considered at the design stage. No methodology currently exists for including this consideration in plant design; its development constitutes a significant research opportunity.

The supercomputer power that is becoming available will provide the opportunity to combine process and control system design—including optimization—into one large problem that can be solved in a way that accounts for their interactions. The success of such a consolidation will depend on the development of approximate compatible models and of techniques to relate model quality to performance.

Robust and Adaptive Control

Control systems are designed from mathematical models that are generally imperfect descriptions of the real process. It is essential that control systems operate satisfactorily over a wide range of process conditions. Thus, the control algorithm must provide for control of the process even when the dynamic behavior of the process differs significantly from that predicted by the model. A control system with this characteristic is sometimes called robust. In fact, a traditional disregard for the model error problem is one of the main reasons for the frequently cited gap between theory and practice in process control. Industry needs algorithms that are robust rather than ones that "get that last half percent performance." Control strategies that work all the time within reasonable limits are better than those that work optimally some of the time but that frequently require reversion to manual control.

Because over time a process often changes, its model parameters must be continually updated; in extreme cases, the basic model must be reformulated. An adaptive system is a control system that automatically adjusts its controller settings or even its structure to accommodate changes in the process or its environment. The problem of model-plant mismatch is of crucial

---

A Revolution in Process Control

Until recently, the chemical process industries have largely tried to solve difficult control problems by using simulation, case studies, and ad hoc on-line adjustment of controller knobs—a procedure that can require personal intuition and continuous human supervision of the controls. These industries did not use modern control techniques because available control theory did not address practical issues. The control theory that was dominant in the 1960s and 1970s assumed that the designer had a linear mathematical model of the process to be controlled as well as a substantial body of fixed knowledge on likely disturbances and the inherent "noise" in the measurements made by process sensors. Such a linear process control theory is built on fatal shortcomings: there are no truly linear processes and there are no fixed, known spectra of information on disturbances and measurement noise. Controllers designed on the basis of linear process theory will fail in the real world of nonlinearity and uncertainty.

New research advances in control theory that are bringing it closer to practical problems are promising dramatic new developments and attracting widespread industrial interest. One of these advances is the development of "robust" systems. A robust control system is a stable, closed-loop system that can operate successfully even if the model on which it is based does not adequately describe the plant. A second advance is the use of powerful semiempirical formalisms in control problems, particularly where the range of possible process variables is constrained.

The development of "internal model control," a design technique that bridges traditional and robust techniques for designing control systems, has provided the framework for unifying and extending these advances. It is now available in commercially available design software.

The combination of these advances is revolutionizing process control, spawning unprecedented research activity in both academia and industry.

---

importance in the design of adaptive controllers for processes since it is that very mismatch that drives changes in the controller parameters. The engineering theory and methodology for designing reliable adaptive controllers for chemical processes are in the earliest stages of development.

Finally, there are always process operations in which neither classical nor modern control is effective. Such operations may require qualitative decision making or the use of past knowl-

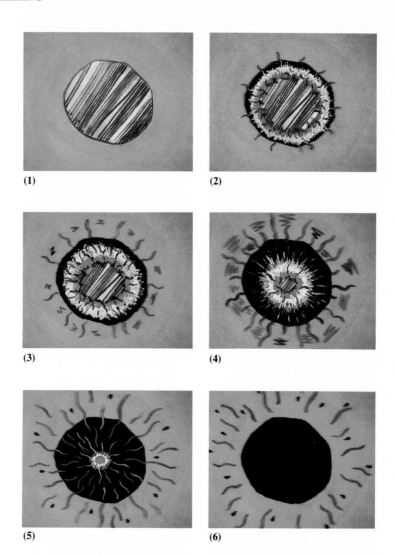

(1)　　　　(2)

(3)　　　　(4)

PLATE 5 Stages in the retorting of an oil shale particle in a hot inert gas are shown. [1, 2, 3] The retorting zone moves inward as products diffuse out of the particle. A coke layer on the particle is formed as a final step of the retorting process. [4, 5, 6] Retorting goes to completion in the center of the particle, and a fully coked particle remains. Courtesy, Amoco Corporation.

(5)　　　　(6)

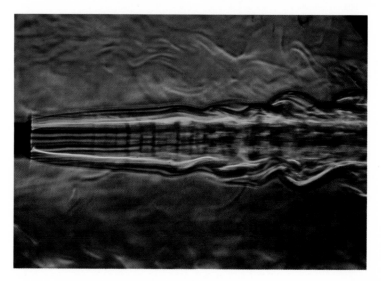

PLATE 6 The turbulent environment in which combustion, a chemical process, takes place is dramatized by this Schlieren photograph of a propane diffusion flame. Courtesy, Norman A. Chigier, Carnegie Mellon University.

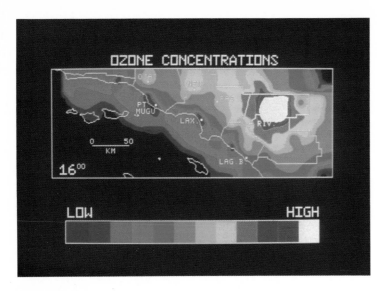

PLATE 7 Chemical engineers develop models to understand the formation, transport, and environmental fate of airborne pollutants such as ozone. This photograph shows a graphic display of a chemical engineering model for ozone concentrations in the Los Angeles basin. Courtesy, John Seinfeld, California Institute of Technology.

PLATE 8 Corrosion is a chemical process whose results are easy to see in the world around us. In this picture, corrosion of reinforcing steel has caused the concrete pillars to spall, weakening the bridge and forcing the installation of wooden joists to temporarily support the bridge deck structure. The results of corrosion impose significant economic costs on society—in 1982, these costs were estimated at about $120 billion. Courtesy, Robert Baboian, Texas Instruments, Inc.

edge. Artificial intelligence techniques offer promise for control system design in these cases.

Batch Process Engineering

The production of fine and specialty chemicals, which are usually made by batch processes, is becoming increasingly important and competitive. The efficient operation of multiproduct and multipurpose batch plants offers a variety of challenging research problems for chemical engineers. Most industrial batch chemical operations are now scheduled by intuitive, ad hoc methods that consist of modest variations around historical operating patterns and that make little or no use of computer technology. It is now widely recognized that the scheduling problems associated with batch processes are immensely complex and, in fact, are among the most difficult combinatorial problems known. Limited progress has been made in using mathematical models in the simplest types of batch process scheduling. Current algorithms are too computationally demanding and complex for industrial use. An important intellectual challenge is to generate a unified field of batch process engineering theory and to put it into a practical context by using case studies.

Linear control theory will be of limited use for operational transitions from one batch regime to the next and for the control of batch plants. Too many of the processes are unstable and exhibit nonlinear behavior, such as multiple steady states or limit cycles. Such problems often arise in the batch production of polymers. The feasibility of precisely controlling many batch processes will depend on the development of an appropriate nonlinear control theory with a high level of robustness.

While startup and shutdown occur relatively infrequently in large continuous plants, they are inherent in batch plant operation. Most startup and shutdown procedures, whether devised empirically or theoretically, are designed to follow a recipe of actions with no feedback. Thus, if upsets occur, there is often no way to change the startup or shutdown in time to avoid unwanted process excursions. Procedures are needed that incorporate feedback and adaptive techniques to the problem of plant startup and shutdown.

PROCESS SENSORS

If we had a completely accurate model of a process and accurate measurements of process disturbances at their inception, then corrective action could be taken directly without the need to measure the output streams from the process after the disturbance has propagated through it. But because we generally do not have adequate models, the output streams of processes must be measured for the purpose of feedback control.

The sensor is the "fingertip" of the process control system. The principal challenge in process sensing is the development of analytical sensors, particularly for determining process stream composition. Such sensors eliminate the need to withdraw samples to determine process and product parameters, a practice that should be minimized because of inherent problems (e.g., samples of reactive intermediates may be toxic or otherwise dangerous, or the intervention represented by withdrawing a sample may affect process operation). Since it is important for process control not to disturb the normal operation of the process, sensors are needed that can operate in the environment of the process stream. The key to meeting this challenge is a fundamental understanding of the physical and chemical interactions at the sensor-environment interface and, in particular, the transport and kinetic processes that occur there.

Future Sensor Developments

The techniques used in the chemical processing of electronic microcircuits (see Chapter 4) are being adapted to the microfabrication of two- and three-dimensional structures for solid-state sensors. These techniques will permit the integration of transducers, optoelectronics, signal-conditioning and data-processing devices, and micromechanical devices into extremely small packages. Reduced size offers advantages in thermal uniformity and response speed; shock and acceleration resistance; and reduced weight, volume, power, and cost.

Solid-state sensors may be developed that will be responsive to a broad range of acoustic inputs, electromagnetic radiation, ionizing radiation, and electrochemical stimuli. Response elements may be tailored for high selectivity among ions, free radicals, or specific compounds. Alternatively, elements with low selectivity are also useful because information from an array of such sensors, each with a different but known broad response, can be processed to provide quantitative analysis of a complex mixture. Complex mixtures also lend themselves to chromatographic analysis. It has been shown that gas chromatographic data can be analyzed on a silicon chip, although with some loss in recognition reliability. Combinations of gas or liquid chromatography or capillary electrophoresis of microsamples with mass spectrometry may be developed to provide superior performance.

The development of biological sensors is taking place at a rapid pace. Biological sensors analyze chemical mixtures using biological reagents of exquisite specificity—for example, enzymes, immunoproteins, monoclonal antibodies, and recombinant nucleic acids. Such sensors may permit the analysis of fast reactions of species in very dilute media. Multicomponent biological sensors may be able to perform complex analyses that involve multiple reactions, with automatic regeneration of the biochemical reagents or removal of interfering species.

Unfortunately, current biological sensors are extremely delicate. Even when the biological reagents are immobilized on a solid carrier, such sensors require careful construction and frequent recalibration, are not always amenable to automation or unattended operation, and sometimes have inconsistent dynamic response and limited life. Although biological reagents are ideally suited for some applications, particularly those in relatively mild environments, they may not survive the harsher conditions often found in process industries. Here again, miniaturization of the biological sensor and its direct integration into an optoelectronic transducer, potentiometric electrode, or membrane are promising approaches. With the current worldwide interest in biotechnology, major innovations in biological sensors can be anticipated.

Further advances in optoelectronics will allow

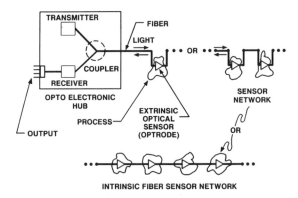

FIGURE 8.4 Configurations for several different kinds of optical fiber sensing systems are shown. The common factor in all these systems is the use of an optical fiber as an integral element in the system, either to carry light to and from discrete sensors (often referred to as optrodes), or as sensitive elements themselves (intrinsic fiber sensors). Courtesy, AT&T Bell Laboratories.

the development of instrumentation with no electrical components in the sensor (Figure 8.4). These devices will operate by transmitting probe light from a remote source to the process sensor with an optical fiber light guide. In the sensor, the light signal will be altered by the sensed environment (e.g., by absorption of certain wavelengths, fluorescence, or scattering) and will thus be "encoded" with information. The encoded signal is transmitted through the optical fiber to a transducer that produces an electronic signal. The advantages of such systems include inherent safety, low signal attenuation, and the ability to multiplex signals in the optical fiber. Such instrumentation can incorporate additional chemical, biological, and electronic components and is likely to play a major role in many future sensor systems.

Future sensors and their associated data processing elements will need capabilities beyond those required for the simple measurement of process variables, such as periodic self-calibration against known standards, automatic compensation for environmental or other interferences, signal conditioning including linearization or other variable calculation, and fault recognition and diagnosis. Some of these capabilities are now available to a limited extent. Others will become available with the continued development of integrated sensors and data processing instrumentation. Arrays of sensors have already been mentioned in connection with complex mixture analysis, but they may also be used to provide redundancy, fault detection, and data reconciliation.

Research Opportunities

The availability of high-quality, real-time information on the conditions and composition of the process stream will permit engineers to develop a completely new generation of process control strategies. The physical, chemical, and biological phenomena at the sensor/process interface must be understood and translated into sensor technology. Chemical engineers are well positioned to contribute to the development of improved process sensors in a variety of ways, including

• work in interdisciplinary collaborations with electronic engineers, biologists, analytical chemists, and others to elucidate the biological, chemical, and physical interactions to be measured;
• application of fundamental principles of reaction engineering and transport phenomena to the design of sensor surfaces;
• development of new process control systems and operation strategies in response to improved capabilities for measurement; and
• determination of the implications for process design of wholly new types of process sensors.

PROCESS ENGINEERING INFORMATION MANAGEMENT

In the next decade, competition among industrialized countries will be influenced by the way in which information and knowledge are managed in industry. The challenge is to be the first to find relevant information, to recognize the key elements of that information, and to apply those elements in the manufacture of desired products.

Computer technology will continue to provide new generations of hardware and software for fast information processing and low-cost storage and retrieval. The use of computers for infor-

mation management and decision making will be essential, and advanced capabilities in user interfaces and networking will bring new dimensions to this application. For example, current on-line literature search systems allow data sharing among many users, significantly increasing individual productivity. However, this technology is generally only used to manage well-organized data; basic research is needed to apply it to engineering data, which are not as well organized.

A process engineer will need to be able to store and access relevant data rapidly in order to carry out process development and design in less time and to solve problems arising from new and complex design requirements (e.g., designing for multiple objectives of profitability, safety, reliability, and controllability). To provide for rapid data storage, access, and transfer, new generations of computer hardware (bulk storage, network, and work stations) and software (data bases) will be needed. Some specific research challenges for chemical engineers, working together with computer scientists and information specialists, follow.

• Most chemical manufacturing processes in the future will be monitored and controlled by computer. Process data will be collected continuously and stored either locally or in a central data base. Research is needed to develop mutually compatible, efficient algorithms for storing and retrieving process data. In addition, computerized procedures will be needed for sifting through voluminous process data for information to use in process improvement and in the generation of new processes.
• Methods must be developed for storing judgments, assumptions, and logical information used in the design and development of processes and models.
• Procedures and methods will be needed for retrieving and operating on other types of imprecise data.
• Efficient transfer of information among engineers and designers will depend on how easily these data can be accessed. The special needs of chemical engineers in this area—the particular ways in which they generate, manage, and use information—merit study.

IMPLICATIONS OF RESEARCH FRONTIERS

The speed and capability of the modern computer, as well as the developing sophistication of chemical engineering design and process control tools, have tremendous implications for the practice of chemical engineering. Chemical engineers of the future will conceptualize and solve problems in entirely new ways. There are two bottlenecks to the application of these powerful resources, though. First, there are not enough active research groups at the frontiers of computer-assisted process design and control. A larger effort is needed in order for the field to keep pace with the expanding power of available computers. Second, many chemical engineering departments lack the computational resources needed to fully integrate advances in design and control into their curricula. For the full potential of the computer to be realized in improved design of chemical products and the improved design and operation of processes to produce them, chemical engineers must be broadly versed in advanced computer technology. This can only happen if they have access to state-of-the-art computational tools throughout their educational careers, not in an isolated course or two.

Making this broad access and utilization of the computer in education possible will require substantial government, academic, and industrial funding to provide both hardware and software. In some cases, groups may need remote access to networks of supercomputers. In other cases, dedicated array processors and other computational hardware may be required in the chemical engineering department. If this country is to maintain a leadership role in chemical technology, critical needs for both research support and facilities acquisition must be addressed. The status of funding in this field and a specific initiative to achieve the goals outlined above are discussed in Chapter 10 and Appendix A.

NINE

Surfaces, Interfaces, and Microstructures

Surfaces, interfaces, and microstructures are key to an improved understanding of catalysis, electrochemistry and corrosion, processes for the manufacture of microcircuits, colloids and surfactants, advanced ceramics and cements, and membranes. Chemical engineers can use their knowledge of thermodynamics, transport phenomena, kinetics, and process modeling to explore a variety of research frontiers. These include the development of molecular-level structure-property relations for guiding the production of materials with specified physical and chemical surface properties; the development of an improved understanding of elementary chemical and physical transformations occurring at phase boundaries; the application of modern theoretical methods for predicting atomic and molecular bonding at surfaces, the energetics and kinetics of surface processes, and the thermodynamics governing the formation of two- and three-dimensional microstructures; and the integration of fundamental knowledge to achieve realistic models of process operation that can be used for process design and evaluation. Chemical engineers will have to integrate their insights with those of researchers in other disciplines (e.g., chemistry and physics) to advance these fields. Access to advanced instrumentation will be critical to the success of chemical engineers in addressing these frontiers.

MICROELECTRONIC CIRCUITS for communications. Controlled permeability films for drug delivery systems. Protein-specific sensors for the monitoring of biochemical processes. Catalysts for the production of fuels and chemicals. Optical coatings for window glass. Electrodes for batteries and fuel cells. Corrosion-resistant coatings for the protection of metals and ceramics. Surface active agents, or surfactants, for use in tertiary oil recovery and the production of polymers, paper, textiles, agricultural chemicals, and cement.

What do these products have in common? They all are based on materials that have precisely defined microstructures and/or surface and near-surface properties. In fact, surfaces, interfaces, and microscale structures are important in virtually every aspect of modern technology; they influence the quality and value of a broad range of products. Modern high-technology materials and products can be thought of as populations of molecules that are distinguished by the ways in which they are spatially organized to provide useful, often unique properties and performance. Organizational forms include microstructures such as domains, mi-

crocrystallites, thin films, micelles, and microcomposites that are assembled into more complex structures on scales from microscopic to larger. The ultimate products formed from microstructures may include tailored forms of particles, fibers, sheets, porous and sponge-like structures, and a vast range of composites and assemblages. So it is that microstructures, interfaces, and surfaces represent an expanding and exciting frontier of untold potential. The frontier extends from humanly designed materials and products all across energy and natural resources processing, environmental engineering, and biochemical and biomedical technologies.

The Nature of Structure

Figure 9.1 illustrates a variety of different structures. This selection is by no means all-inclusive; a host of related structures such as colloids, microstrands, thin films, microporous solids, microemulsions, and gels could also have been shown. The parts of each of these structures are distinguished by the zones—interfaces—between them, which often seem to be

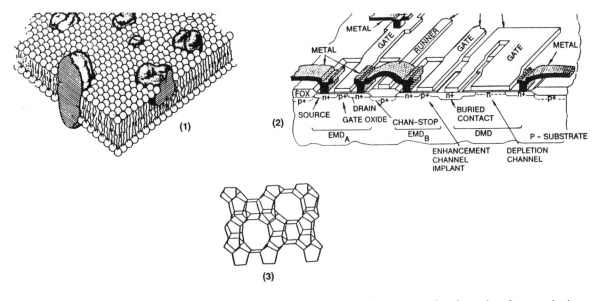

FIGURE 9.1 Examples of different types of surfaces, interfaces, and microstructures. (1) A biological membrane composed of phospholipid molecules in which protein molecules are embedded. (2) An NMOS logic circuit. (3) A section of ZSM-5 zeolite.

so sharply defined that they are called surfaces. Well-defined surfaces occur between solids and either gases or liquids and thus are commonly found in catalytic and electrode reactions. More diffuse interfaces may occur between solids, as in microelectronic devices, and between fluids or semifluids, as in many polymeric and colloidal systems.

Whether they are called surfaces or interfaces, when the zones between parts of a structure are "thin"—from a fraction of a micrometer (the limit of the ordinary microscope) down to molecular dimensions—the matter in them assumes a character that is somewhat different from that seen when the same matter is in the form of a bulk solid. This special character of a molecular population arranged as an interfacial zone is manifested in such phenomena as surface tension, surface electronic states, surface reactivity, and the ubiquitous phenomena of surface adsorption and segregation. And the structuring of multiple interfaces may be so fine that no part of the resulting material has properties characteristic of any bulk material; the whole is exclusively transition zones of one kind or another.

The finer the scale of structuring in a material, the more the material is taken up by interfacial zones. As the proportion of material in interfacial zones increases, the special character of those zones begins to dominate the properties of the total structure. This is why the performance of highly microstructured materials is often marked by superior physical properties (e.g., mechanical strength, toughness, and elasticity) or chemical properties (e.g., resistance to oxidation or corrosion, selective permeability, and catalytic activity). Further, since "the performance is in the interface," there are strong economic incentives to develop high-performance products from inexpensive bulk materials, by modifying their surface and interfacial properties, to compete with existing products composed of expensive, homogeneous materials.

Relationship to Applications of Chemical Engineering

Surface and interfacial properties and processes affect virtually every aspect of modern

chemical engineering. The ability to produce microstructures with the desired properties is leading to a wide variety of new materials and products that promise to improve our quality of life and to provide new business opportunities for U.S. industry. Five specific examples of the societal and technological impacts of microstructural engineering, corresponding to Chapters 3 through 7 in this report, are given in the following subsections.

Biochemical and Biomedical Engineering

Many biological processes depend on cellular microstructure. These include selective and reaction-coupled transport, antibody-antigen interactions, enzyme catalysis, and the synthesis of proteins and membranes. Surface and interfacial phenomena affect cell growth through their influence on cell immobilization, cell-cell interactions, and cell disruption and separation of constituents. A wide variety of therapeutic products, then, exert their influence on living systems by influencing molecular events occurring at biological interfaces. In addition, the practical implementation of cell culture for the commercial production of biochemicals (biotechnology) is heavily dependent on advances in understanding how cell-surface interactions mediate important cellular events. Finally, surface and interfacial phenomena are important to the function of a variety of biomedical devices, including artificial tissues and organs, sustained-release drug delivery systems, and future generations of biosensors.

Electronic, Photonic, and Recording Materials and Devices

Very-large-scale integrated (VLSI) microelectronic circuits epitomize intricately designed solid microstructures that are painstakingly built under meticulously controlled conditions. Structure scales in microelectronic devices are now at the 1 μm-level and are expected to reach the 1-nm level in the next decade. The production of such devices depends on the carefully controlled deposition, patterning, and etching of thin layers of metals, semiconductors, polymers, and ceramics. The growing field of optoelectronics depends on the formation of carefully tailored glass films and fibers to serve as light guides. Recording materials also depend on the careful generation and control of microstructure. Thus, magnetic memories require ultrafine particulate coatings, whereas the new field of optical storage requires polymeric and inorganic thin films that undergo finely tuned structural changes upon illumination. The production of these and many other materials will present a growing set of challenges to chemical engineers in the future.

Polymers, Ceramics, and Composites

Microstructures, surfaces, and interfaces play a central role in the production of ceramics, glasses, and organic and inorganic polymers. The mechanical and chemical properties of cement and concrete are also highly dependent on the formation of the proper microstructure. Interfacial chemistry is critical in determining the strength and durability of fiber- and fabric-reinforced composites and laminated high-performance polymer composites. Exciting opportunities are now emerging for molecular composites of rod- and coil-type polymers.

Processing of Energy and Natural Resources

The recovery of petroleum from sandstone and the release of kerogen from oil shale and tar sands both depend strongly on the microstructure and surface properties of these porous media. The interfacial properties of complex liquid agents—mixtures of polymers and surfactants—are critical to viscosity control in tertiary oil recovery and to the comminution of minerals and coal. The corrosion and wear of mechanical parts are influenced by the composition and structure of metal surfaces, as well as by the interaction of lubricants with these surfaces. Microstructure and surface properties are vitally important to both the performance of electrodes in electrochemical processes and the effectiveness of catalysts. Advances in synthetic chemistry are opening the door to the design of zeolites and layered compounds with tightly specified properties to provide the de-

sired catalytic activity and separation selectivity.

Environmental Protection, Process Safety, and Hazardous Waste Management

The production of aerosols, soot, ash, and fines during combustion, calcining, and incineration depends on a large number of physical and chemical processes at gas-solid and/or gas-liquid interfaces. Similar phenomena are important in the formation of suspensions, slimes, sludges, slurries, and other waterborne particulates from natural resource processing and all sorts of chemical manufacture. The separation of particulates from air or water requires highly microstructured separators (e.g., membranes, colloidal adsorbents, porous adsorbents, and micellar and reverse-micelle scavengers). Likewise, the understanding of soil contamination and decontamination by hazardous wastes depends on knowing soil and mineral microstructure as well as the interactions of waste materials with these porous matrices.

INTELLECTUAL FRONTIERS

Crucial to the future of chemical engineering are two profound questions that define the field's frontiers:

• How do the properties of a system or product—and thus its processing characteristics—depend on the local structure of the system (i.e., the size and shape of its parts, their contacts and connectivity, and their composition)?

• How does local structure depend on the starting materials and processing conditions by which the system or product was created? How should the process be designed and controlled to achieve reliably the desired structuring?

The answers to both questions can come only from interdisciplinary research. The first question impels physicists, chemists, materials scientists, geologists, biologists, and engineers alike. So does the second, but it serves as a more potent driving force to materials engineers and process engineers. If chemical engineers do

FIGURE 9.2 This high-resolution electron micrograph shows the unique pore structure of the ZSM-5 zeolite catalyst. Molecules such as methanol and hydrocarbons can be catalytically converted within the pores to valuable fuels and lubricant products. Courtesy, Mobil Research and Development Corporation.

connected porespace filled with fluid). But as the dimensions of the supporting matrix become smaller, access to the catalyst on the matrix surface becomes more strongly controlled by diffusion, a process quite slow compared to convection, which is favored by larger interstices and pores. Moreover, the activity and specificity of the catalyst itself are often influenced by the way the catalyst is deployed on the supporting surface, by the nature of that surface, and by the catalyst's interactions with the surface.

The development of new and improved catalysts requires advances in our understanding of how to make catalysts with specified properties; the relationships between surface structure, composition, and catalytic performance; the dynamics of chemical reactions occurring at a catalyst surface; the deployment of catalytic surface within supporting microstructure; and the dynamics of transport to and from that surface. Research opportunities for chemical engineers are evident in four areas: catalyst synthesis, characterization of surface structure, surface chemistry, and design.

not take up the research opportunities that come to them in the area of structure, researchers in other disciplines will. But chemical engineering has much to contribute to interdisciplinary attacks on structure-function relationships and process-structure connections. These potential contributions, and the research opportunities associated with them, are discussed in the remainder of this section.

Catalysis

Catalysts are most often used to promote reactions of fluid reactants. They are, with some exceptions, colloidal, amorphous, or microcrystalline states that, to be accessible by the reactants, are deployed on supporting matrices with a large ratio of surface area to volume. The largest possible ratios are achieved, though, by suspending fine particles containing the catalyst in the fluid of reactants, thus creating the problem of removing the fine catalytic particles from the fluid after the reaction is complete. Alternatively, the fine particles may be packed together in one of several ways (e.g., as a porous bed; as the internal surface of a fine, consolidated porous medium; as an interspersion of connected solid; or as a semisolid and

Catalyst Synthesis

The introduction of new types of catalytic materials has often led to the development of new or improved chemical processes. Examples are zeolite catalysts for petroleum cracking and organic synthesis (Figure 9.2), platinum-based reforming catalysts, Ziegler-Natta catalysts for olefin polymerization, and catalysts for control of automobile exhaust emissions. Synthetic inorganic chemistry is currently opening up ways of preparing new multicomponent compounds, many of which have compositional and geometrical characteristics that suggest they might have potential as catalysts. Such materials include the molecular sieves based on silicon aluminum phosphates, pillared clays and other layered materials, supported transition metal clusters, and metal carbides and nitrides. There is a growing interest in studying the influence of promoters on the catalytic properties of nonnoble metals, which could lead to a reduction in the demand for catalysts based on metals such as platinum, palladium, and rhodium—metals for which the United States is almost totally dependent on foreign sources. Catalysts that do not contain these metals but possess many of their catalytic properties have recently been developed.

A challenge particularly suited to chemical engineers is the development of process models for predicting the microstructure and surface structure of catalysts as a function of the conditions of their preparation. Such models could be used not only to guide the preparation of existing materials, but also to explore possibilities for making novel catalysts.

Characterization of Catalyst Structure

Characterization of catalyst structure and composition is essential to achieving a fundamental understanding of the factors controlling catalyst activity, selectivity, and stability. During the past 15 years, the application of surface science techniques (e.g., low-energy electron diffraction [LEED], Auger electron spectroscopy [AES], x-ray photoelectron spectroscopy [XPS], and ultraviolet photoelectron spectroscopy [UPS]) has led to a very rapid advance in understanding how metallic catalysts function. Knowledge of the

Zeolites and Shape-Selective Catalysis

The technological progress of the petroleum refining industry has been marked by the discovery and development of ways to turn more of a given barrel of crude oil into products such as gasoline, diesel fuel, and heating oil. These three products contain collections of hydrocarbon molecules that lie within certain size and molecular weight limits. Converting fractions of the barrel containing molecules outside these ranges into molecules in the useful range is accomplished, in part, by a series of reactions known as cracking.

Cracking has long been carried out using solid acid catalysts such as silica-alumina. The discovery of Faujasite-type zeolite catalysts (see illustration) in the 1950s led to greatly improved cracking catalysts for gasoline production. These catalysts did not change the chemistry of cracking, but markedly improved selectivity for gasoline products due to their pore-related control of key hydrogen-transfer reactions. With these catalysts, the nation's gasoline needs have been supplied from significantly less crude oil, saving the economy billions of dollars over the last 15 years.

Another set of constituents in crude oil that can cause problems in petroleum products is the long, higher-boiling straight-chain hydrocarbons. They can crystallize in petroleum, causing it to lose fluidity at room temperature and above. The fluidity of diesel fuel and heating oil can be markedly improved by removing these waxy hydrocarbons. Typical cracking catalysts cannot accomplish this task because of the low reactivity of straight-chain hydrocarbons. A different type of zeolites, though, can help. They are the smaller-pore ZSM-5 zeolites (see illustration). These zeolites can discriminate between molecules based on their size and shape, a property known as reactant shape selectivity. Only those molecules that are small enough to enter their pores are cracked, and straight-chain hydrocarbons insert easily. Thus, ZSM-5 zeolites can be used to selectively dewax petroleum feedstocks, increasing the amount of diesel fuel that can be produced from a barrel of crude oil by up to 10 percent. Lubrication oils can also be dewaxed by shape-selective cracking of the waxy molecules. This elegant and inexpensive approach to dewaxing is practiced in refineries worldwide.

Shape-selective zeolites can also be used to discriminate among potential products of a chemical reaction, a property called product shape selectivity. In this case, the product produced is the one capable of escaping from the zeolite pore structure. This is the basis of the selective conversion of methanol to gasoline over

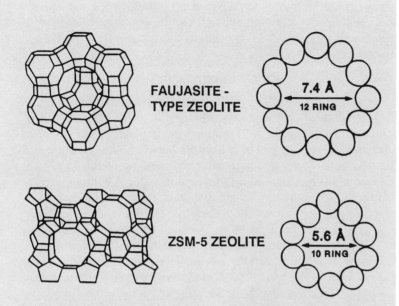

FAUJASITE - TYPE ZEOLITE

ZSM-5 ZEOLITE

7.4 Å
12 RING

5.6 Å
10 RING

The framework structures and pore cross-sections of two types of zeolites are shown. (Top) A Faujasite-type zeolite has a three-dimensional channel system with pores of at least 7.4 Å in diameter. A pore is formed by 12 oxygen atoms in a ring. (Bottom) ZSM-5 zeolite has interconnected channels running in one direction, with pores 5.6 Å in diameter. ZSM-5 pores are formed by 10 oxygen atoms in a ring. Reprinted with permission from *Chemical Engineering Progress*, 84(2), February 1988, 32.

ZSM-5, the process by which one-third of New Zealand's gasoline is produced. Related processes for the production of ultrahigh quality jet and diesel fuel, olefins, and *p*-xylene are based on the same zeolite and the same principles.

Many chemical reactions, especially those involving the combination of two molecules, pass through bulky transition states on their way from reactants to products. Carrying out such reactions in the confines of the small tubular pores of zeolites can markedly influence their reaction pathways. This is called transition-state selectivity. Transition-state selectivity is the critical phenomenon in the enhanced selectivity observed for ZSM-5 catalysts in xylene isomerization, a process practiced commercially on a large scale.

Zeolite chemistry is an excellent example of how a three-dimensional surface can alter the course of chemical reactions, selecting for one product out of a host of potential candidates. In addition to the many commercial applications that they have found, shape-selective zeolites have provided the basis for a rich new area of catalytic science and technology, one expected to spawn yet more materials, knowledge, and applications.

structure of supported metal catalysts has been advanced through the use of high-resolution transmission electron microscopy (TEM), extended x-ray absorption fine structure spectroscopy (EXAFS), and, more recently, solid-state nuclear magnetic resonance spectrometry (NMR). Major advances in understanding zeolite structure have resulted from combining information obtained from x-ray diffraction, silicon-29 and aluminum-27 NMR, infrared spectroscopy, and neutron diffraction. Unfortunately, many of the currently known techniques must be used ex situ, making it difficult to observe catalyst structure and composition during use or to examine the dynamics of the changes in these properties. Therefore, it is essential that greater attention be given to developing in-situ characterization techniques based on infrared spectroscopy, Raman spectroscopy, EXAFS, NMR, and neutron diffraction.

Surface Chemistry

Knowledge of the structure and bonding of molecules to surfaces has been obtained from such techniques as LEED, electron energy-loss spectroscopy (EELS), secondary-ion mass spectrometry (SIMS), infrared spectroscopy (IRS), Raman spectroscopy, and NMR spectrometry. The scope of such studies needs to be greatly expanded to include the effects of coadsorbates, promoters, and poisons. Greater emphasis should be given to developing new photon spectroscopies that would permit observation of adsorbed species in the presence of a gas or liquid phase. Attention also needs to be given to studies of surface reaction dynamics to obtain a fundamental understanding of the elementary reaction steps involved in the decomposition or rearrangement of an adsorbed species and of its reaction with coadsorbed species to form either desirable or undesirable products. Knowledge gained from such studies combined with information on the structure of the catalyst surface will lead to an improved understanding of what types of centers are critical for achieving high activity and selectivity and the role of poisons and other substances in causing catalyst deactivation. The information gained from such studies will provide vital input to large-scale scientific computations of molecular dynamics aimed at predicting the influence of surface composition and structure on catalyst performance.

Catalyst Design

Recent theoretical studies have demonstrated that it is possible to calculate accurately adsorbate structure and energy levels, to explain trends with variations in metal composition, and to interpret and predict the influence of promoters and poisons on the adsorption of reactants. Additional efforts along these lines will contribute greatly to understanding how catalyst structure and composition influence catalyst-adsorbate interactions and the reactions of adsorbed species on a catalyst surface. With sufficient development of theoretical methods, it should be possible to predict the desired catalyst composition and structure to catalyze specific reactions prior to formulation and testing of new catalysts.

Electrochemistry and Corrosion

Electrochemical processes (e.g., electrolysis, electroplating, electromachining, current generation, and corrosion [Plate 8]) are distinguished by their occurrence in a boundary region between an electrolyte (liquid or solid) and an electrode. The course of these processes is strongly dependent on the potential at the electrode surface, the composition and structure of the electrode, the composition of the electrolyte, and the microstructure of the electrolyte in the boundary layer near the electrode surface. In certain applications, the pore size and connectivity of the electrode can also be important.

Charge Transfer

There are two issues of fundamental importance to the kinetics of electrochemical processes. The first is the dependence on distance of electron transfer between sites that are not in contact. An understanding of this is critical for creating three-dimensional catalytic structures through which charge percolates to fixed sites. The second is the kinetics of electron transfer at well-defined sites such as individual defects on single crystals or on selected planes of carbon. An improved understanding of the physical processes governing charge percolation and conduction through an electrode and the factors influencing electron transfer at the electrolyte-electrode interface is needed. Such knowledge would make it possible to choose electrode materials and tailor their microstructure to suit particular applications. The identification of cheap electrode materials to replace platinum would be a very significant accomplishment.

Molecular Dynamics

Mechanistic studies are needed on a select number of electrochemical reactions, particularly those involving oxygen. These studies are far from routine and require advances in knowledge of molecular interactions at electrode surfaces in the presence of an electrolyte. Recent achievements in surface science under ultrahigh vacuum conditions suggest that a comparable effort in electrochemical systems would be equally fruitful.

There is no fundamental theory for electrocrystallization, owing in part to the complexity of the process of lattice formation in the presence of solvent, surfactants, and ionic solutes. Investigations at the atomic level in parallel with studies on nonelectrochemical crystallization would be rewarding and may lead to a theory for predicting the evolution of metal morphologies, which range from dense deposits to crystalline particles and powders.

Many electrochemical reactions consist of

Fuel Cells for Transportation

Imagine if we could extract significantly more useful energy out of our precious fuel resources! Think how remarkable it would be to carry out combustion processes at efficiencies not possible in even the most sophisticated heat engines. These are not empty dreams. Such a device was first demonstrated in 1839. Called a fuel cell, this electrochemical device may eventually reshape major energy use patterns throughout society.

Many of the early giants of chemistry were fascinated by the potential of fuel cells. Their early attempts to use fuel cells to "burn" coal failed, though, because of materials problems and operating difficulties. It is only recently that advances in science and engineering have resulted in practical and reliable fuel cells. Today, NASA's space shuttle relies on high-pressure hydrogen-oxygen fuel cells to generate electrical power while in orbit. But how can we bring the potential of fuel cells back to earth?

If fuel cells could be used in transportation vehicles, it could have a major impact on worldwide consumption of petroleum. Major improvements that are needed for this to happen include increasing the efficiency of fuel cells, increasing their power density, reducing their manufacturing cost, and developing fuel cell designs capable of rapid start-up.

Surface and interfacial engineering are the key to addressing these technological needs. The conversion of chemical energy to electrical energy in a fuel cell takes place at the interface of an electrode with an electrolyte. Inexpensive materials are needed for electrodes—materials that combine long-term stability with the ability to catalyze the conversion reaction. Inexpensive electrolytes are needed with molecular structures that favor the transport of protons and oxygen to reactive sites at the interface. The prospects for finding superior materials will be improved by developing new techniques for in-situ characterization of solid-liquid interfaces.

There is exciting new engineering and science in the old field of fuel cells. Even more important, the payoff of this research may be a truly stunning advance in our use of energy.

rosion, electrodeposition, or etching. Mechanistic details remain essentially unknown. Improved insight would benefit technologies that depend on the formation and stabilization of controlled surfaces.

Supramolecular Microstructures

As noted earlier, the kinetics of electrochemical processes are influenced by the microstructure of the electrolyte in the electrode boundary layer. This zone is populated by a large number of species, including the solvent, reactants, intermediates, ions, inhibitors, promoters, and impurities. The way in which these species interact with each other is poorly understood. Major improvements in the performance of batteries, electrodeposition systems, and electroorganic synthesis cells, as well as other electrochemical processes, could be achieved through a detailed understanding of boundary layer structure.

Since electrochemical processes involve coupled complex phenomena, their behavior is complex. Mathematical modeling of such processes improves our scientific understanding of them and provides a basis for design scale-up and optimization. The validity and utility of such large-scale models is expected to improve as physically correct descriptions of elementary processes are used.

complex sequences of steps, such as in electroorganic synthesis. In these, the key to high yield is knowledge of the sequence so that adroit choices of electrode and solution materials can be made. More thorough documentation of rate and equilibrium constants is mandatory to transfer such scientific understanding into engineering practice.

The influence of added agents and inhibitors is important in processes that involve cor-

Electronic, Photonic, and Recording Materials and Devices

The creation of microstructure with well-defined electrical or optical properties is critical to the production of integrated circuits and recording materials. The processes used to define microstructure in this context are described in Chapter 4. Common to the fabrication of all electronic and photonic materials are the deposition, patterning, and etching of thin films. The preparation of magnetic recording materials also involves the creation of very small magnetic particles and their distribution in a thin layer of binder on a substrate. Virtually all aspects of these processes involve surface and interfacial phenomena. The challenge to chemical engineers is to understand the fundamental elements of each processing step at a level where this knowledge can be used to guide the design and fabrication of high-density, superfast circuits and storage devices.

The scientific problems that must be addressed to meet the challenges posed by the decreasing feature and domain size include the following:

• characterization of microstructures;
• identification of the factors affecting the controlled application and development of photoresists;
• determination of the elementary processes involved in chemical vapor deposition, plasma deposition, and etching of thin films; and
• mathematical modeling of all aspects of microstructures formation (e.g., in photoresist spincoating, resist patterning, and thin film deposition and etching).

Characterization of Microstructure

Advances in integrated circuit technology and in the production of high-density storage devices depend on making ever-smaller microstructures. An essential aspect of this problem is the ability to characterize the physical and chemical properties of domains having dimensions between 0.1 and 1 μm. Visualization and elemental mapping of microstructural elements at this scale have been accomplished by use of scanning electron microscopes and, more recently, high-resolution scanning Auger and x-ray photoelectron spectrometers. If chemical engineers are to play an effective part in the future of the electronics and photonics industries, they must be familiar with such modern analytical devices.

Photoresist Processing

Polymer films that are sensitive to light, x-rays, or electrons—known as photoresists—are used extensively to transfer the pattern of an electronic circuit onto a semiconductor surface. Such films must adhere to the semiconductor surface, cross-link or decompose on exposure to radiation, and undergo development in a solvent to achieve pattern definition. Virtually all aspects of photoresist processing involve surface and interfacial phenomena, and there are many outstanding problems where these phenomena must be controlled. For example, the fabrication of multilayer circuits requires that photoresist films of about 1-μm thickness be laid down over a semiconductor surface that has already been patterned in preceding steps.

A planar resist surface is essential to the successful execution of subsequent steps, but it is as yet difficult to attain. A knowledge of the ways in which polymer viscoelasticity, surface tension, and surface adhesion affect the rheology of resist flow is needed. Another area requiring research is the solvent development of resist films after exposure of the films to radiation through a mask. In this step, it is essential to remove only those parts of the polymer that have been degraded by the radiation. Research is needed to understand how solvent composition, residual polymer stress, polymer adhesion, and the swelling of an unirradiated polymer affect the geometric definition of the developed well. Such problems become increasingly important as resolution is pushed to 1 μm and then to 0.1 μm.

Chemical Vapor Deposition and Plasma Deposition/Etching of Thin Films

Both thermal and plasma-assisted chemical vapor deposition techniques are used routinely to deposit thin films, and plasma etching is used to define fine features in the films. Understanding the fundamental reactions involved in these processes is essential to developing an understanding of how best to control the deposition or etching of thin films and the design of equipment to carry out such steps. To make progress in this area, chemical engineers need to identify the chemical species present in the gas phase of both thermal and plasma reactors through the use of such techniques as emission spectroscopy, laser-induced fluorescence spectroscopy, electron spin resonance spectroscopy, and mass spectrometry. The dynamics of gas-phase chemical reactions need to be understood for processes involving not only neutral species but also electrons and ions. Another task of equal importance is understanding how reactive gas-phase species interact with solid surfaces to achieve film growth or etching. While some of the elementary processes are similar to those occurring at the surface of catalysts, others, such as ion bombardment and photon-assisted etching, are specific to systems found in the electronics and photonics industries. Because of their knowledge of transport phenomena, chemical engineers are expected to contribute significantly to an understanding of how local electrical fields and concentration gradients interact to influence such processes as the anisotropic etching of semiconductors.

Mathematical Modeling

The systems involved in microelectronic processing are usually so complex that they can rarely be described by simple conceptual models. It is therefore necessary to develop mathematical models that incorporate fundamental information in order to understand such processes adequately. The advantage of mathematical modeling has been demonstrated for simple systems, and more detailed models will continue to appear with the growing access to large-scale computing. The conventional macroscopic models will have to be augmented with microscopic treatments of interface formation so that process conditions and interface properties can eventually be related. A close collaboration between experimentalists and theoreticians will lead to detailed models for simulating such processes as chemical vapor deposition, plasma etching, photoresist spinning, and photoresist development.

Colloids, Surfactants, and Fluid Interfaces

The area of colloids, surfactants, and fluid interfaces is large in scope. It encompasses all fluid-fluid and fluid-solid systems in which interfacial properties play a dominant role in determining the behavior of the overall system. Such systems are often characterized by large surface-to-volume ratios (e.g., thin films, sols, and foams) and by the formation of macroscopic assemblies of molecules (e.g., colloids, micelles, vesicles, and Langmuir-Blodgett films). The peculiar properties of the interfaces in such media give rise to these otherwise unlikely (and often inherently unstable) structures.

The formation of ordered two- and three-dimensional microstructures in dispersions and in liquid systems has an influence on a broad range of products and processes. For example, microcapsules, vesicles, and liposomes can be used for controlled drug delivery, for the containment of inks and adhesives, and for the isolation of toxic wastes. In addition, surfactants continue to be important for enhanced oil recovery, ore beneficiation, and lubrication. Ceramic processing and sol-gel techniques for the fabrication of amorphous or ordered materials with special properties involve a rich variety of colloidal phenomena, ranging from the production of monodispersed particles with controlled surface chemistry to the thermodynamics and dynamics of formation of aggregates and microcrystallites.

The current status and the emerging opportunities in the science of colloids, surfactants, and fluid interfaces can be addressed conveniently by considering a threefold hierarchy of systems as follows:

- individual molecules (e.g., surfactants),
- self-assembling (associated) microstructures of surfactants and other molecules in the colloidal size range; and
- macroscopic (often structured) systems made up of associated microstructures and bulk phases.

This last category may also include fluid interface systems with unstructured bulk phases and/or moderate surface-to-volume ratios.

Significant breakthroughs have been made in recent years in the identification, preparation, characterization, and understanding of entities at all these levels, creating new opportunities for the successful use of colloidal and interfacial phenomena in chemical engineering and presenting new challenges as described below.

New surfactant molecules are being designed

with novel and special properties, particularly with the inclusion of fluorine atoms and silicon-containing substituents (both yielding surface activity in organic media). Multifunctional surfactants are being designed as coupling agents, release agents, rewetting agents, and steric stabilizers for wet colloids. Other recent developments include the synthesis of di-tail (and even tri-tail) surfactants for use in the preparation of thin films and membranes. Despite these advances, a host of possibilities for structural modifications of hydrophilic and hydrophobic groups in surfactants remains unexplored.

Self-assembling structures include monolayers and micelles, both of which have received much study. However, new approaches and possibilities for these structures are emerging. For instance, chromophores are being incorporated in monolayer assemblies to produce Langmuir-Blodgett films with a variety of unique optical properties. There has been great interest in incorporating chemical functionality into monolayer-forming surfactants to permit lateral polymerization, either in monolayers on liquid substrates or in Langmuir-Blodgett films on solids, thus yielding exceedingly thin films or membranes with structural integrity. For micelles, greater refinement in the determination of micellar shapes, structures, and properties, as well as the investigation of the kinetics of micelle formation and disintegration have become possible thanks to recent advances in the use of photon correlation spectroscopy, small-angle neutron scattering, and neutron spin-echo spectroscopy. Notable advances have also been made in the study of

Flat-Panel Image Displays: An Emerging Application for Chemical Engineering Principles

The most common imaging device that we encounter is probably the picture tube in a television set. It's a bulky device and it requires a lot of electrical power to operate. These two disadvantages have sparked the search for a better alternative—a display with the imaging power of the picture tube, but one that is more portable and that has lower power requirements.

One candidate for the "picture tube of the future" is the electrophoretic image display, or EPID. EPIDs are flat-panel displays that contain submicron particles of pigment suspended in a liquid along with a dye that provides contrast. When an electrical potential is applied to the system, pigment particles are driven to the interface between the suspending liquid and a viewing plate (usually glass). There they can be seen under normal illumination. EPIDs have the potential of providing an image that has extremely high optical contrast under normal light conditions, that is legible over a wide range of viewing angles, that is inherently retained on the display (as opposed to an image needing constant refreshing as on a TV picture tube), and that requires low voltage and power.

The EPID concept can be combined with other developments in imaging technology to produce optical devices such as light valves and x-ray imagers. The accompanying illustration shows an example of an electrophoretic x-ray imager.

The most crucial part of this chemical display technology is the liquid dispersion of pigment particles and dye. The dispersion must be stable, even when the particles are compressed to form an image where the concentration is 10 times higher than in the bulk of the liquid. The pigments must be able to retain their charges after numerous switching operations. The fluid must allow fast response time. Unwanted migration of particles when an electrical field is applied and other electrohydrodynamic effects must be controlled.

Many of the crucial problems for researchers in this area are the same as the ones encountered in other areas of surface and interfacial science. The research of chemical engineers on high-performance ceramic materials, field-induced bioseparations, and fouling also addresses phenomena such as agglomeration and clustering in dispersions and rheology of dispersions. For EPIDs,

other microstructural assemblies, and new entities have been discovered and identified. These include inverse micelles, vesicles, liposomes, bilayers, microemulsions, liquid crystallites, and a variety of as yet unnamed entities formed by the interaction of dissolved polymers (often

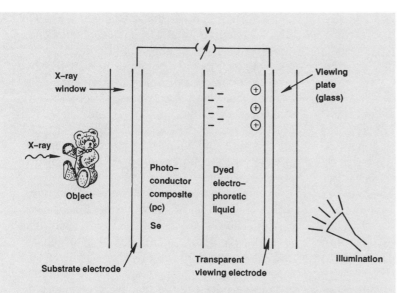

Large-area solid-state x-ray receptor with an electrophoretic image display. When a voltage is applied across the image cell, pigment particles and counterions in the electrophoretic liquid separate. Most of the applied voltage drop occurs across the Se layer. X-ray exposure under this condition leads to the creation of a charge image at the pc-liquid interface due to the generation of x-ray-induced charges in the Se. After the x-ray exposure, the applied voltage is reduced to zero, and the pigment particles are driven to the viewing plate. The image becomes visible upon illumination.

some research questions where chemical engineering principles are particularly relevant include the following:

• How do particles interact when they are repeatedly and forcefully packed and unpacked in electrical fields?

• How are particles, counterions, polymers, and surfactants transported across an EPID cell?

• How does fluid motion in an EPID device affect this overall transport?

• How fast do structures formed by charged particles dissipate in applied fields?

Questions similar to these are being researched by chemical engineers who are active in the broad field of colloid and interface science. Future contributions by chemical engineers may be the key to the success of the EPID concept in visual display applications.

Methods to determine and control the properties of individual surfactant molecules and to determine the conditions needed to produce well-defined molecular assemblies are just beginning to emerge. We are at the threshold of being able to produce deliberately structured supramolecular entities with properties tailored to meet special applications. Some additional examples of problems that will have significant impacts over the next one or two decades follow:

• New methods to produce large quantities of mono-sized particles of nearly any inorganic material desired (e.g., metals, oxides, silicates, sulfides) are needed for the processing of ceramics, electronic materials, and other engineered materials.

• New methods of emulsion polymerization, particularly the use of swelling agents, to produce monodisperse latexes of any desired size and surface chemistry are also needed. Perfect spheres as large as 100 μm can now be produced in the zero-gravity environment provided by the space shuttle. These spheres and other mono-sized particles of various shapes can be used as model colloids to study two- and three-dimensional many-body systems of very high complexity.

• Refinements in the theory of interparticle long-range van der Waals forces (the Landau-Lifshitz theory) are within reach. New techniques are now available for measuring the complex dielectric constants of various media required for the implementation of that theory.

• Recognition and description of new interparticle interaction forces such as those owing to magnetic dipoles, steric and electrosteric repul-

proteins) or other macromolecules with the above structures. Many of these exist in the cellular makeup of living tissues (their study is called membrane mimetic chemistry), and host-guest systems or artificial enzymes may also be produced.

sion, and long-range solvent ordering offer opportunities to study interfacial molecular phenomena that were previously difficult to describe.

• New experimental techniques for the direct measurement of interparticle forces are now available and can be used to understand the physicochemical factors that control adhesion, coating phenomena, tribology, and others.

• New optical (static as well as dynamic) techniques for the study of long-range order in structured continua are beginning to appear and can be used to understand the constitutive properties and relations in complex (polymeric, nematic, and other structured) fluids.

• New application of modern statistical mechanical methods to the description of structured continua and supramolecular fluids have made it possible to treat many-body problems and cooperative phenomena in such systems. The increasing availability of high-speed computation and the development of vector and parallel processing techniques for its implementation are making it possible to develop more refined descriptions of the complex many-body systems.

• Because of the increasing level of control that is now possible in the preparation of model colloids and surfactants, model many-body systems can be created in the laboratory and studied by non-intrusive instrumental techniques in parallel with computational and theoretical sophistication.

In view of the above developments, it is now possible to formulate theories of the complex phase behavior and critical phenomena that one observes in structured continua. Furthermore, there is currently little data on the transport properties, rheological characteristics, and thermomechanical properties of such materials, but the thermodynamics and dynamics of these materials subject to long-range interparticle interactions (e.g., disjoining pressure effects, phase separation, and viscoelastic behavior) can now be approached systematically. Such studies will lead to significant intellectual and practical advances.

Ceramics, Cements, and Structural Composites

The development and control of microstructure are critical in the processing of ceramics and cements. The chemical engineer's knowledge of reaction kinetics, surface phenomena, and transport phenomena could contribute effectively to the development of new materials.

Bulk ceramics are produced conventionally by the sintering of powders. The strength, toughness, thermal stability, and dielectric properties of the fired ceramic depend strongly on the size and uniformity of the precursor powder and on the chemical properties of the powder surface.

Examples of the need for improved ceramics technology, either to produce ceramics more economically or to produce ceramics with improved performance, abound in both structural and electronic applications. They include automobile engine components, armor, welding nozzles, artificial hip joints, wear elements of valves and pumps, cutting tools, electronic packages, and a host of other current or future applications of this exciting new area of materials science. Among new developments in ceramic materials are ceramic glasses, microporous ceramic filtering media, ceramic-ceramic composites and microcracked composite ceramics for catalyst supports, ceramic fibers, ceramic thin films and coatings, permselective membranes for application in separations and sensors, and a variety of high-performance cements.

Essential to these improved ceramics is the control of particle size and uniformity through well-characterized chemical reactions. Chemical engineers have rich opportunities for contribution through surface and interfacial engineering of preceramic particles and powders. Specific research areas include the study of chemical reactions affecting powder particle nucleation, precipitation, surface structure and composition, size distribution, shape, shape distribution, surface charges, agglomeration, deagglomeration, tribological characteristics, and rheology.

Important research opportunities in surface and interfacial engineering also exist with respect to the properties of finished ceramic bodies, such as surface energy and susceptibility to crack propagation. Sintering mechanisms and kinetics represent a very important area for scientific investigation. Progress in addressing

High-Tech Concrete

We are surrounded by concrete, from our basement floors to the support structures of many office buildings, from the sidewalks outside to dams and bridges all across America. Because concrete is so inexpensive and commonplace, it is not often thought of as a high-technology material. But research into chemical and mineral additives is yielding new insights into the complex chemical reactions that occur when concrete sets. These insights, in turn, are leading to improvements in the inner structure of concrete. By finding ways of decreasing concrete porosity and increasing the strength of its interparticle bonds, researchers have pushed the maximum compressive strength of concrete upwards from 4,000–6,000 psi to more than 40,000 psi.

The secret of the new high-performance concrete is the use of chemicals to modify surfaces and interfaces inside the bulk material. The best-known chemical additives are potent surfactants, the so-called superplasticizers and high-range water reducers, that reduce overall porosity in fresh concrete while maintaining its workability prior to hardening. The addition of highly divided minerals, such as microsilicon, in combination with a superplasticizer greatly improves concrete properties such as abrasion resistance and resistance to concrete breakdown by deicing salts, seawater, and freeze-thaw cycles.

The payoff to society from greater attention to the surface and interfacial engineering of concrete is potentially immense: high-tech concretes that will prolong the life of public works and reduce their maintenance costs as well as dramatic new applications for this old reliable material.

steps and can be influenced by a host of low-concentration additives. Examples include the superplasticizers, which not only reduce the viscosity of freshly mixed concrete but also affect its final microstructure. Collaborative research among chemical engineers, civil engineers, and colloid and surface chemists can accelerate progress toward achieving superior formulations for cement and concrete.

Membranes

Membranes are thin two-dimensional structures designed to pass preferentially certain components. Highly efficient separations of gaseous or liquid components can be achieved with such technologies as reverse osmosis, ultrafiltration, gas separation, microfiltration, dialysis, and electrodialysis. In these systems, separations are driven either by pressure or by a gradient in chemical or electrochemical potential. Membranes are also finding increasing use in controlled drug release devices and biosensors. Traditional applications of membrane technology have barely scratched the surface of an exciting and rapidly developing area.

There are two major frontiers in membrane research, one technological and the other scientific. At the technological frontier, chemical engineers can make important contributions to the development of new materials, the engineering of structure or morphology into membranes, and the identification of new ways of using permselective membranes.

On the materials side, there is considerable interest in developing novel membrane materials that are functionalized to selectively adsorb a specific component from a fluid phase. Membrane materials that are environmentally stable and resistant to fouling are also needed. Since

these issues may permit the application of ceramics of known high-performance characteristics to areas in which their use is now uneconomical.

Finally, there is a need for chemical engineers to bring their expertise in surface and interfacial engineering to the problems of developing better varieties of Portland cement and concrete. Both these commodities are produced in vast quantities each year, and major improvements in their properties (e.g., freeze-thaw durability, corrosion resistance, and compressive strength) would tremendously benefit society. The properties of Portland cement and concrete are controlled by the microstructure of the materials. The microstructure of concrete is developed through a remarkably complex series of higher fluxes of permeates can be achieved by decreasing membrane thickness, there is increasing emphasis on building structural integrity into the membrane. Possibilities include the use of laminated polymer membranes and porous ceramic substrates for ultrathin polymer layers.

On the applications side, integrated membrane processes represent an attractive area to be developed, in which a membrane separation is combined with a conventional separation to accomplish a job neither process by itself could do. An example is the distillation of azeotropic mixtures, where a pervaporation module can be used to get around the azeotropic composition. Another example is the use of hollow fiber membranes in bioreactors. Here the membrane acts as a support for an immobilized cell or enzyme and at the same time facilitates the supply of oxygen and/or nutrients. In a particularly elegant extension, a permselective membrane may be combined with a catalytic membrane to selectively remove a dilute reactant from a stream containing inerts and to generate a product stream in which the product concentration is many-fold higher than that of the reactant going in. Finally, there are some very exciting opportunities for the development of "smart" membranes that respond to the types or concentrations of species present in the fluids contacting them. One example is a membrane that regulates the delivery of insulin from a reservoir into a patient's bloodstream in response to blood sugar level.

The scientific base for rationally designing membrane polymers for specific applications is very limited, and hence there is an immense frontier to be conquered. Work is also needed on transport fundamentals, structure-permeability relationships, and elucidation of how to control membrane morphology. While phenom-

Modern Membranes: Process Technology Enables New Product Design

How do new materials and processing technologies affect product design? Recent developments in membrane-based products show how design strategy has been profoundly changed by the development, by chemical engineers, of new materials and processes for spinning hollow fibers.

Ten years ago, most membranes were made in flat sheets or by assembling numerous sheets into sandwiches. This quasi-two-dimensional geometry imposed significant restrictions on membrane product design. For example, sandwiches of planar membranes were difficult to engineer because of the problem of sealing a multiplicity of thin, flat components around their edges. The advent of materials and processing technology for spinning hollow fibers, a chemical engineering achievement, has changed the way that membranes find use in product design.

Hollow tubular membranes have important physical advantages. If the tube is small in diameter (and about 1 mm is typical), most polymers can withstand considerable "hoop stress," i.e., pressure differential across the wall. A tube with a small diameter also has a high ratio of surface area to volume. This provides a large amount of interfacial area, a desirable characteristic in a membrane that operates by passing substances through its skin selectively. By tailoring the composition of the hollow fiber surface, one can vary the kind of separation it will perform. If the wall of the fiber is made from a microporous material, ultrafilters can be made that will separate cells from a fermentation broth or a single size-fraction of proteins from a mixture. If the wall of the fiber is made

enological transport models already exist, molecular-scale models for describing the transport of organic permeants and the transport of condensible vapors through glassy or nonequilibrium matrices have yet to be developed. The application of structural probes, such as carbon-13 NMR spectroscopy and XPS, could contribute to the development of structure-permeability probes. Likewise, elucidation of the physical and chemical processes involved in membrane synthesis could aid in producing membranes with the desired microstructures.

RESEARCH NEEDS

To understand how the properties and performance of a material are tied to its microstructure and how microstructure depends on proto allow only one kind of molecule to pass through, one can make membranes such as Monsanto's Prism® membranes that separate hydrogen from other gases or GE's oxygen enrichment membranes.

The future is bright for devices based on hollow-fiber membranes. By combining, in one device, membrane layers whose properties change in response to their environment, one will be able to make self-regulating or "smart" membranes. An example might be a hypothetical artificial pancreas that would regulate the passage of insulin to a diabetic. A two-layer hollow tube, capped at both ends, would encapsulate a reservoir of insulin. The outer layer of this device would have the enzyme glucose oxidase immobilized within it. When glucose in the bloodstream reacts with this enzyme, hydrogen peroxide is released. The hydrogen peroxide would diffuse into the second layer, a redox membrane whose ability to pass insulin out from the central reservoir is controlled by the amount of hydrogen peroxide diffusing into and out of this layer. The overall system would release insulin in proportion to the amount of glucose in the bloodstream, just like the pancreas.

Techniques such as enzyme immobilization and the production of layered structures of redox polymers are already used in the development of diagnostic sensors for clinical and biological applications. To make devices like the artificial pancreas a reality, these techniques must be combined with hollow fiber technology to engineer membranes with the appropriate interfaces. It is an area where chemistry, physics, biology, and engineering interact. Chemical engineers have the holistic perspective needed to gather these strands together.

cessing, researchers must be able to detect microstructure, characterize it, resolve its shape and connectivity, and measure its size and composition. They need to visualize the microstructure, whether directly through some sort of microscopy or indirectly by means of theory based on model-dependent synthesis from measurements. The challenges are enormous because of the small size and complexity of microstructures, the fluidity and thermal fluctuations of liquid and semiliquid systems, and the rapidity of many physical transformations and chemical reactions.

Instrumentation

Instrumentation for experimental observation and measurement is paramount in microstruc-

ture-related research. One reason that surfaces, interfaces, and more complicated microstructures are a frontier of chemical engineering and processing research is that modern science has recently spawned a number of microstructural probes of unprecedented resolution and utility. For the first time, we have the proper tools to attack the molecular and chemical basis of microstructures.

Of course, our understanding of microstructures will be advanced through an interplay of observation, conceptualization, experiment, and theory. But in this area of engineering science, advances will come most often when already developed instrumental probes are adapted to new systems or new probes are perfected to answer questions arising from practical problems. The adaptation and development of instrumental probes for systems of interest to chemical engineers demand cooperative efforts with the originating scientific disciplines and with instrument manufacturers. In such efforts, chemical engineers can bring important refinements or innovations to instrumental practice. This has already occurred, for example, in the development of video-enhanced optical microscopy, rapid-freezing cryo-electron microscopy, the analysis of solid catalytic surfaces, and the probing of solid-liquid interfaces important in electrochemical catalysis.

Microscopy and Microtomography

The direct visualization of microstructure may be accomplished by various forms of microscopy. Recent refinements in microscopy techniques are epitomized by video-enhanced interference phase-contrast microscopy, which is emerging as a workhorse probe for colloidal suspensions and other microstructured liquids.

This technique is capable of resolving structures at distances approaching the wavelength of visible light (350 to 800 nm).

Another useful tool, and arguably the most powerful probe of surface topography on scales from those of the light microscope down to 5 nm, is scanning electron microscopy with x-ray microanalysis. This technique combines magnifying power, depth of field, and ability to analyze local composition. It may also be used to study the internal microstructure of specimens by fracturing them (sometimes after freezing). Scanning electron microscopy is certain to become a very useful tool in the hands of chemical engineers, particularly as they apply the principles of chemical engineering science (e.g., a sophisticated understanding of heat and mass transfer, phase change, and chemical reactions) to interpreting images and developing ancillary techniques.

Even greater magnifying power is provided by transmission electron microscopy, and in some instances this technique can be complemented with energy loss spectroscopy. Transmission electron microscopy can resolve microstructure down to atomic scales (0.1 to 0.5 nm) and requires the skillful application of specialized techniques to extremely thin, solid, or solidified specimens (or replicas of specimen surfaces such as internal surfaces of fractured samples). Correct interpretation of images requires not only considerable experience, but also a fundamental understanding of sample behavior during preparation and under the electron beam and of the contrast mechanisms underlying an image.

Scanning tunneling microscopy is a recent invention of great potential (Figure 9.3). Capable of resolving surface topography down to atomic dimensions, it operates perfectly well on surfaces immersed in gas or liquid, whereas electron microscopy requires that the specimen be studied under a vacuum (except for special

FIGURE 9.3 Measuring less than 1/100,000,000th of an inch, the hills in this micrograph are individual atoms on a silicon crystal that have been enlarged more than 1 billion times using a scanning tunneling microscope. The microscope collects digital information that is plotted by a computer. Bands on the hills are contours assigned by the computer to help researchers see how crystalline structures are formed. Copyright AT&T, Microscapes.

"environmental stages" that function only with severely reduced effectiveness). However, the intense electrical fields of the scanning tip can strongly affect the specimen locally. Equipment and techniques are rapidly being refined, and it appears that scanning tunneling microscopy will be playing an important role as a probe of active solid-gas and solid-liquid interfaces.

X-ray microtomography is a new development of great promise for reconstructing, displaying, and analyzing three-dimensional microstructures. Resolution of around 1 μm has been demonstrated with currently available synchrotron sources of x-rays, x-ray detectors, algorithms, and large-scale computers. The potential for microstructural research in composites, porous materials, and suspensions at this and finer scales appears to be tremendous.

Magnetic resonance imaging, or microtomography by multinuclear magnetic resonance, is another new development that is even more exciting because it provides three-dimensional mapping of the abundance of a variety of atoms. Compositional aspects of microstructure can thereby be resolved. However, the resolution of currently available instruments does not yet approach 1 μm.

Scattering Methods

Beams of electromagnetic radiation of appropriate wavelength are scattered when they interact with the gradients inherent in structured materials. By measuring the ways in which the intensity of scattered radiation varies as a function of the angle at which the radiation initially strikes the sample, the wavelength of the radiation, and the time, many aspects of the structure of materials can be inferred.

Bulk heterogeneities and surface topography are both marked by electron distributions that vary in their polarizability. These variations are capable of scattering photons. In liquid and semiliquid materials, where the variations themselves fluctuate over time and space, static light scattering and its dynamic complement—photon correlation spectroscopy—are important probes of larger colloidal-scale microstructures and their thermal motions, which are often fingerprints of structure. For solids, the scattering of x-ray radiation can be used to characterize the structure of both crystalline and amorphous materials. Of particular interest in terms of amorphous materials is the technique of extended x-ray absorption fine structure, which provides information on atomic coordination number and local bond distances.

Generally, the more intense the available beam source, the shorter the time scales, the weaker the heterogeneities, and the longer the distances that can be probed by a scattering method. Hence, there is a strong drive to utilize high-powered lasers, synchrotrons, and intense neutron sources in research on surfaces, interfaces, and microstructures. This is particularly true in the study of liquid materials and of systems that undergo rapid physical transformations or chemical reactions.

Resonance Spectroscopies

The interaction of radiation with a material can lead to an absorption of energy when the radiation frequency matches one of the resonant frequencies of the material. The exact frequency at which the absorption occurs and the shape of the absorption feature can provide detailed information about electronic structure, molecular bonding, and the association of molecules into microstructural units.

Nuclear magnetic resonance (NMR) spectroscopy is an enormously powerful tool that, in chemistry, has become a mainstay for analyzing molecular structure and environment. In recent years, NMR spectroscopy has proved useful for studying catalysts, amorphous semiconductors, and colloidal-scale microstructure and molecular aggregates. Examples of the application of NMR spectroscopy to problems of interest in chemical engineering include identification of the secondary building units involved in zeolite synthesis, analysis of the development of bicontinuous "liquid microsponge" in surfactant-oil-water systems, the clustering of hydrogen in amorphous silicon photovoltaic devices, and the structural characterization of carbonaceous deposits that lead to formation of coke on catalysts. In addition to providing time-averaged information, NMR spectroscopy can be used to probe the dynamics of molecular motion on time scales ranging from 10^6 to 1 second. Thus, for example, time-resolved NMR techniques have made it possible to characterize the dynamics of forming the precursors to zeolite synthesis.

The vibrations of molecular bonds provide insight into bonding and structure. This information can be obtained by infrared spectroscopy (IRS), laser Raman spectroscopy, or electron energy loss spectroscopy (EELS). IRS and EELS have provided a wealth of data about the structure of catalysts and the bonding of adsorbates. IRS has also been used under reaction conditions to follow the dynamics of adsorbed reactants, intermediates, and products. Raman spectroscopy has provided exciting information about the precursors involved in the synthesis of catalysts and the structure of adsorbates present on catalyst and electrode surfaces.

Molecular-level characterization of surface composition and structure can be obtained through a variety of electron and ion spectroscopies. The two-dimensional structure of surfaces and ordered arrays formed by adsorbates is revealed by low-energy electron diffraction

(LEED). This technique can also be used to follow phase changes and surface reconstruction in real time. The atomic composition of surfaces can be determined by Auger electron spectroscopy (AES), x-ray photoelectron spectroscopy (XPS), and secondary ion mass spectrometry (SIMS). While SIMS provides the highest elemental sensitivity, AES and XPS can resolve spatial variations in composition down to 0.1 μm. XPS, in addition, gives information on the valence state of individual atoms, from which details of interatomic bonding can often be inferred. The density of electrons in bonding orbitals can be obtained from ultraviolet photoelectron spectroscopy (UPS). When carried out with monochromatic beams of synchrotron radiation, this technique can also identify the orientation of individual atomic and molecular orbitals at a solid surface. With the exception of LEED, each of these techniques can be used to characterize polycrystalline films, amorphous materials, and powders, as well as single crystals.

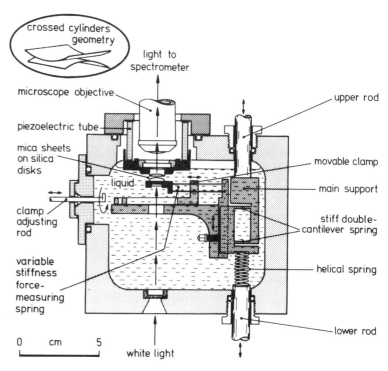

FIGURE 9.4 The direct force measurement apparatus shown here can measure the forces between two curved molecularly smooth surfaces in liquids. Mica surfaces, either raw or coated, are the primary surfaces used in this apparatus. The separation between the surfaces is measured by optical techniques to better than 10 nm. The distance between the two surfaces is controlled by a three-stage mechanism that includes a voltage-driven piezoelectric crystal tube supporting the upper mica surface; this crystal tube can be displaced less than 10 nm in a controlled fashion. A force-measuring spring is attached to the lower mica surface and its stiffness can be varied by a factor of 1,000 by shifting the position of a movable clamp. Reprinted with permission from *Proc. Natl. Acad. Sci. USA*, 84, July 1987, 4722.

Other Important Methods

The statics and dynamics of microstructures are governed by the forces that create or maintain them. Rarely can the forces be measured directly. But forces between special surfaces immersed in fluid can now be accurately gauged at separations down to 0.1 nm with the direct force measurement apparatus, an ingenious combination of a differential spring, a piezoelectric crystal, an interferometer, and crossed cylindrical surfaces covered by atomically smooth layers of cleaved mica (Figure 9.4). This recent development is finding more and more applications in research on liquid and semiliquid microstructures, thin films, and adsorbed layers.

Its use will continue to expand as its cost falls, its complexity decreases, and its capabilities multiply.

The electron tunneling microscope tip is currently gaining recognition as the most exquisite micromanipulator for measuring local deformability of solid surfaces, down to nanometers and smaller. For microstructures on scales of micrometers and larger, micromanipulation apparatus from biology and biophysics is turning up in probes of deformability and force; microelectronic devices are in the offing. Microelectrode probes continue to evolve. Laser-doppler motion probes capable of micrometer resolution and birefringence and dichroism measurements are becoming important in the characterization of many surfaces.

Particulate microstructures, as well as the fragments obtained by disrupting more extensive structures, are separated by equipment of varying cost and sophistication: ultrafilters, ultracentrifuges, gel permeation and size exclusion chromatographs, and electrophoretic separators. The ultimate goal is rapid, automatic sorting of individuals from a population of particles upon sensing one or more properties of each. In the laser spectroscopy cell sorter, this capability has reached down to the scale of living cells through application of several technologies, including ink-jet printing. The cell sorter has already opened up research into the population structure of cell cultures, a basic research problem important in biotechnology.

Cost and Availability

Instruments for probing microstructures and their changes typically follow a rule that the costs of purchase, installation, operator expertise, and equipment maintenance become higher as the dimensions of the structure to be measured become smaller, as the connectivity and shape to be examined become more complex, and as the time between events to be resolved becomes shorter. Management of such research gets complicated as its scale moves into the gray area between small science and big science. The need arises to share instruments within a department, an institution, or a regional center, or, in some cases, a national or international facility, an activity that can become cumbersome when the instruments being shared are central to an investigation. Cutting-edge research often calls for improvement, adaptation, and augmentation of equipment. Research can be stultified when scheduling a shared instrument inhibits hot pursuit of a finding or an idea, when the needs of others prevent modification of an instrument, or when a key experiment faces the risk of a temporary shutdown of an instrument.

Over the long run, the costs of sophisticated equipment fall to the level where the equipment can be acquired for use and adaptation by individual research groups. But in the interim, while costs are sufficiently high that the level of usage of advanced equipment does not justify widespread acquisition, or while the requisite financial resources cannot be marshaled, an effective strategy may be for funding agencies or institutions to provide reasonably complete subsets of the sophisticated instruments needed for surface and microstructural engineering to selected groups that would focus on a coherent theme. The result would be to create a small community of users (e.g., three to four faculty members) with similar interests in terms of use of the equipment. While some needs of the group may go unmet, requiring the use of equipment in other locations, the investigators should generally be able to carry out the major portion of their preparative and analytical work in close proximity to their laboratories.

Theory

Significant advances are needed in our current understanding of how molecules interact with surfaces and with each other to form microstructural units. Theoretical efforts along these lines should be carried out starting at the molecular level and extending to the level of bulk materials. The development of a hierarchy of theoretical methods for predicting the behavior of increasingly complex ensembles of molecules will be invaluable in understanding how best to process materials. Examples of specific areas requiring development were discussed earlier in this chapter.

Development of the necessary theoretical models will involve a careful integration of insights from different disciplines. Concepts new to chemical engineers (e.g., fractals, Monte Carlo methods, and percolation theory) will have to be introduced to provide more accurate and/or computationally efficient means for formulating process descriptions. Chemical engineers will need to become more familiar with recent advances in applied mathematics and computer science in order to work productively with researchers from these disciplines. In particular, collaborative efforts between theoreticians and experimentalists should be encouraged as a means to new theoretical approaches and insights.

The need for access to supercomputers, discussed in detail in Chapter 8, cannot be over-emphasized. In the past, many of the major problems in the processing of structured materials have yielded to analysis once sufficient computational power was provided to permit the utilization of very detailed physical models. Supercomputers have made possible significant advances in the modeling of plasma reactors, complex electrochemical systems, coating flows, and stress fracturing of polymers and ceramics. Advanced computational tools will become even more important as chemical engineers attack the important and highly complex problems now on the cutting edge of research on surfaces, interfaces, and microstructures.

IMPLICATIONS OF RESEARCH FRONTIERS

There is an increasing societal need for materials with surface and interfacial properties tailored to meet specific application. This spectrum of materials is extremely broad; it ranges from thin films for microelectronic circuits, to high-strength concrete for roads and buildings, to membranes for food protection. The development and production of such advanced materials, and of surface active agents, will be rich in technical challenges for chemical engineers.

To address these challenges, chemical engineers will need state-of-the-art analytical instruments, particularly those that can provide information about microstructures for sizes down to atomic dimensions, surface properties in the presence of bulk fluids, and dynamic processes

with time constants of less than a nanosecond. It will also be essential that chemical engineers become familiar with modern theoretical concepts of surface physics and chemistry, colloid physical chemistry, and rheology, particularly as it applies to free surface flow and flow near solid boundaries. The application of theoretical concepts to understanding the factors controlling surface properties and the evaluation of complex process models will require access to supercomputers.

Funding must be provided to support research at academic and industrial institutions. Researchers in universities will require funds for research assistants, instrumentation, computer time, and travel to use special facilities such as synchrotron radiation sources, neutron sources, and atomic resolution microscopes. The primary support for these efforts should come from federal agencies, with additional support provided by industry. Industry will also need to finance its own research and development efforts. One should anticipate that generic long-range work will be carried out at universities, whereas research leading to specific products and processes will be conducted primarily in industrial laboratories. Collaborative investigations between university and industry scientists should be strongly encouraged, since such efforts will help define the goals and objectives of intermediate- and long-range research and facilitate the transfer of new ideas and techniques into practice.

TEN

Recommendations

THIS REPORT HAS been about the future contributions of chemical engineering research to the national well-being, and about emerging research frontiers of special importance. Some of the implications of these frontiers have already been explored in individual chapters. In this chapter the committee draws together its principal recommendations for action by policymakers in four sectors: academia, industry, government, and professional societies.

RECOMMENDATIONS FOR ACADEMIA

Education and Training of Chemical Engineers

Chemical engineering undergraduate curricula have traditionally been designed to train students for employment in the conventional chemical processing industries. The current core curriculum is remarkably successful in this effort. Chemical engineers will continue to play a major role in the chemical and petroleum industries, but new areas of application as well as new emphases on environmental protection; process safety; and advanced computation, design, and process control will require some modifications of the curriculum.

These modifications will not entail a massive revision of the curriculum. Continued emphasis is needed on basic principles (e.g., thermodynamics, transport, separations, reaction engineering, and design) that cut across many applications. A new way of teaching these principles,

though, is needed. Students must be exposed to both traditional and novel applications of chemical engineering. Emerging applications should be highlighted and an expanded range of process scales and chemistries should be taught.

Chemical engineers of the future will not only be interested in the transport and chemical reactions of small molecules, they will also be interested in large organic molecules, proteins, polymers (both organic and inorganic), interfaces, surfaces, solids, and composite materials. Each of these new areas of interest—to the extent that they are not now treated in depth in the core courses of the curriculum—will require new emphasis in teaching. Core courses such as separations (unit operations) should be modified to cover new challenges. For example, a separations course might be modified to address challenges in bioprocessing and ultrapurification by placing less emphasis on distillation, absorption, and conventional extraction, and more emphasis on examples of generating high selectivity in separations.[1] Similarly, unit operations laboratories and design courses in most institutions are in need of rethinking and revitalization to accommodate new needs and opportunities.

Another important need in the curriculum is for a far greater emphasis on design and control for process safety, waste minimization, and minimal adverse environmental impact. These themes need to be woven into the curriculum wherever possible. The AIChE Center for Chemical Process Safety is attempting to provide curricular material in this area, but a larger effort than this project is needed. Several large chemical companies have significant expertise in this area. They should do more to share their insights and methods with academic researchers and educators. (See "Recommendations for Industry" below.)

Certain emerging specialties of chemical engineering will require a deeper exposure of engineering students to other disciplines. For example:

• Students interested in biochemical engineering will need to become familiar with the language and concepts of the biosciences, particularly molecular biology.
• For those interested in advanced materials, more in-depth knowledge of statistical and continuum mechanics than that provided in transport and mechanics courses seems essential.
• Chemical engineering students interested in electronic materials need to understand the elements of electrical engineering and solid-state physics in order to work productively with colleagues in these disciplines.
• A greater emphasis on surface and interfacial phenomena is needed for all chemical engineers interested in materials engineering.

It will not be easy to shoehorn all of these elements into a 4-year program, and the initiation of carefully delimited electives and options for students is probably the best approach. The committee applauds the recent action of the AIChE to allow more flexibility in the choice of science electives by undergraduates in accredited departments. The development of such electives and new course sequences is not a casual undertaking. It will require greater involvement by faculty in providing advice and counsel to students at the sophomore and junior level, who may not be prepared to make definitive choices about specialization in their major.

The mix of industries into which chemical engineers are being introduced is changing rapidly. Chemical engineers have always been proud of their flexibility and willingness to rise to new challenges. This characteristic will be needed more than ever in the future, as diversity within the processing industries greatly increases. For example:

• There will be a greater variety of products and processes, increased demands for quality, and a necessity to shift rapidly from one product to another.
• Chemical engineers will be working with a broader spectrum of colleagues; not just chemists, but physicists, biologists, material scientists, and electrical engineers.
• The chemical processing industries will increasingly have a global outlook. They will require engineers with a greater understanding of the global economy, knowledge of engineering research and advances occurring in other countries, and appreciation for other cultures.
• These industries will be operating in a complex regulatory environment. They will require engineers with a broad perspective on economic and environmental policy.
• With the current emphasis on streamlined management and lean staffing in the processing industries, chemical engineers will have fewer peers and fewer superiors; their ability to make decisions will be tested early and often.

Chemical engineering faculty need to consider these challenges in planning for tomorrow's more broadly based curriculum.

The Future Size and Composition of Academic Departments

How can chemical engineering departments implement this broader curriculum as well as aggressively respond to new research challenges and opportunities? A bold step is needed. The committee recommends that universities conduct a one-time expansion of chemical engineering departments over the next 5 years, exercising a preference for new faculty capable of research at interdisciplinary frontiers.

This expansion will require a major commitment of resources on the part of universities, government, private foundations, and industry. Can such a preferential commitment to one discipline be justified, particularly at a time of severe budgetary austerity? Each agency responsible for funding chemical engineering research will have to answer this question for itself, but the committee believes that the following arguments for expansion are worthy of consideration.

• U.S. leadership in technology is under challenge today as never before. Future markets for biotechnology products, advanced materials, and advanced information devices will be won by those who are best able to integrate innovative product design with efficient process design. Chemical engineers, the uniquely "molecular" engineers, have powerful tools that can be refined and applied to these challenges. They can make significant contributions to U.S. capabilities in these high-technology areas.

• The existing chemical processing industries are high-technology enterprises too. Moreover, the chemical industry is one of the most successful U.S. businesses on world markets, with a $7.8 billion trade surplus in 1986. The chemical processing industries require a continuing supply of highly trained engineers armed with the latest insights from the core research areas of chemical engineering. Redirecting resources from these established and vital research programs to the emerging applications of the discipline risks reducing the U.S. leadership position in a large and very successful industrial sector. Is this an appropriate course to take at a time when industrial competitiveness and the large U.S. trade deficit are such pressing national concerns?

Increasing the number of research groups and the resource levels in chemical engineering seems to be the most practical way of (1) moving chemical engineers aggressively into the new areas among this report's research priorities while (2) maintaining the discipline's current strength and excellence. The following recommendations for universities are intended to ensure that such an additional investment would be used most effectively.

• Chemical engineering departments should use additional resources to appoint and encourage faculty to address the frontier areas identified in this report. Many of these frontiers will require more emphasis on interdisciplinary research. The selected appointment of chemical engineering faculty whose backgrounds are from

other disciplines should be encouraged. Cross-disciplinary research partnerships between chemical engineers and colleagues in other disciplines should be stimulated and rewarded. (See sections on "Cross-disciplinary pioneer awards" and "Cross-disciplinary partnership awards" under "Recommendations for Government.")

• Academic institutions should make substantial commitments (e.g., cost sharing and maintenance for instrumentation and facilities) to make additional outside investments in chemical engineering as effective as possible. This will be particularly needed to justify additional funding for advanced instrumentation and computers needed for research and education.

• Academic institutions should take steps to strengthen even further the long-standing ties between chemical engineering departments and the chemical process industries, with a particular emphasis on the flow of personnel between academia and industry. Mechanisms for increased interchange include appointing industrial researchers as adjunct professors in academia; providing industrial sabbaticals for university professors; inviting industrial researchers as seminar speakers on campus; placing student researchers in industrial laboratories; and expanding the number of masters degree programs, evening continuing education programs, and other short courses targeted to industrial research personnel.

• Chemical engineering departments should use additional resources to promote greater interaction with small process technology firms in electronics and biotechnology. These firms play an unheralded but key role in developing industrial technology in these areas. (See section on "Improvment of Links Between Chemical Engineering Departments and Small Process Technology Firms" under "Recommendations for Government.")

RECOMMENDATIONS FOR INDUSTRY

The committee encourages decision makers in the chemical processing industries to increase their commitment to research in their companies both to retain their world leadership in chemical technology and to deliver to American society the maximum benefit from its investment in basic research. These industries should take advantage of opportunities to leverage their funds by working cooperatively in generic non-proprietary areas, such as in-situ processing and solids processing (see Chapter 6) and health, safety, and environmental protection (see Chapter 7).

Industry should also continue to commit resources to academic research, for reasons that go far deeper than the desirability of additional funds. The development of any engineering field, and particularly one as closely linked to manufacturing as chemical engineering, needs the intellectual guidance that can only come from an industry with a stake in research outcomes. Also, industry has to be linked to academia so that new laboratory results can be rapidly transferred to product and process design. An industry committed to financial sponsorship and personnel exchanges with academia will make sure that the crucial industrial intellectual involvement needed for success exists. Thus, the committee urges that:

• industry continue to expand its support of academic chemical engineering research in each of the priority research areas identified in this report, and
• policymakers in industry facilitate the exchange of people across the interface between industry and academia. A number of possible mechanisms for such exchanges are listed under "Recommendations for Academia." They will be most effective if corporate culture and incentive structures promote their use.

One area where intellectual involvement from industry is essential is in the area of process safety. Several large chemical companies have committed substantial resources to advancing their understanding of this area. The generic insights that have emerged from their work need to be shared with academic researchers and educators as well as with smaller chemical companies. The AIChE Center for Chemical Process Safety is one mechanism for such transfer of knowledge. Others, including direct contacts between academic and industrial researchers in this area, are needed.

RECOMMENDATIONS FOR GOVERNMENT

Balanced Portfolios

Maintaining the health of any research enterprise is a difficult undertaking. It requires the recognition that future research advances are unpredictable and may substantially reorder the initial priorities assigned to particular areas of investigation. It demands an appreciation that researchers in different areas of science and technology, or at different stages in their professional careers, have different requirements for support. To maintain the vitality of chemical engineering, and to enable its researchers to make the greatest possible contribution to solving the problems facing society, a philosophy of support for the field is needed that can be described in terms of three "balanced portfolios": one of priority research areas, one of funding sources, and one of support mechanisms.

The committee's balanced portfolio of priority research areas has already been discussed in the Executive Summary and the preceding chapters of the report. A discussion of the portfolio of funding sources is presented later in this chapter and, more comprehensively, in Appendix A. The following section discusses the committee's views on a balanced portfolio of support mechanisms.

Support Mechanisms in Perspective

Different research frontiers require different mixes of support mechanisms. The appropriate mix for a particular area depends on several of factors, including the nature of the scientific area; its requirements for expensive equipment, instruments, or facilities; and the need for trained personnel from that area in the broader economy.

Table 10.1 summarizes a range of possible funding mechanisms for research. Many of these mechanisms are well known and need no further elaboration. Some of them, however, are new mechanisms being proposed by the committee and are briefly described below. The great number of possible support mechanisms for academic research should be placed in the

TABLE 10.1 Possible Support Mechanisms for Chemical Engineering Research

Support Mechanism	Likely Applicants
Single Investigator Awards	
Starter grants	New investigators
Presidential Young Investigator awards	Extremely promising new investigators
Solicited project grants	Younger investigators
Career development awards	Extremely meritorious younger investigators
Unsolicited project grants	Meritorious midstream investigators
Research excellence awards[a]	Extremely meritorious midstream and senior investigators
Cross-Disciplinary Awards	
Cross-disciplinary pioneer awards[a]	Promising new investigators moving into chemical engineering from other disciplines
Cross-disciplinary partnership awards[a]	Two or three co-principal investigators from different disciplines
Equipment and Facilities	
Equipment and instrumentation grants	One or more research groups with special requirements
Regional and national equipment facilities	Researchers needing unique equipment or facilities (e.g., synchrotrons or neutron-scattering facilities)
Centers and Academic-Industrial Consortia	
NSF Science and Technology Centers	Collaborative groups of principal investigators
Engineering Research Centers	Cross-disciplinary engineering research attacking several aspects of a common problem area
Other centers and consortia[a]	Large research centers funded by industry at academic campuses
Improving links to small high-technology firms[a]	Small process technology companies working with academic investigators

[a] New mechanism discussed in the text.

following perspective. Academic chemical engineering research has traditionally taken place in the milieu of the small research group. Led by a single principal investigator, the small research group enjoys the advantages of decentralization and freedom to move in new directions as opportunities unfold. The environment of a small group, where individual students take responsibility for significant parts of the group's work, also facilitates the training of independent-minded researchers. Much of the vitality of chemical engineering over its history can be traced to innovative and exploratory research groups led by individual principal investigators.

Because of these advantages, and because many of the most exciting research problems in chemical engineering lend themselves to the scale and environment of the small research group, grants to individual academic investigators should continue to remain the mainstay of the portfolio of support mechanisms for research.

In this context, how can interdisciplinary research on the frontiers of the discipline be best conducted? There is an obvious role for large assemblages of researchers in centers of various types. The committee also believes that the special needs of interdisciplinary research areas can be met by creating two new funding mechanisms modeled on the best features of the small group model.

• Cross-disciplinary partnership awards. The first mechanism is designed to encourage two or three small research groups to submit joint applications. Such a partnership would allow researchers to begin to communicate across disciplinary lines without necessitating the creation of large and formal entities. Cross-disciplinary partnership awards should be considered in a variety of research areas where interdisciplinary cooperation is vital, and particularly for research areas associated with the emerging technologies described in Chapters 3 through 5 of this report. A case in point would be biochemical engineering where more effective links in research and training are needed between chemical engineers and life scientists.

• Cross-disciplinary pioneer awards. The second proposed interdisciplinary funding mechanism would encourage top-quality young researchers trained in other disciplines (e.g., molecular biology, chemistry, materials science, and solid-state physics) to accept tenure-track faculty positions in chemical engineering departments at leading universities. The committee believes that these departments would be enriched by the judicious appointment of young faculty who are at the cutting edge of disciplines important to chemical engineering frontiers. Some appointments of this type are already being made, but these "pioneers" face considerable obstacles in making the transition from the area in which they were trained to another discipline. These include the difficulties of immersing oneself in a new intellectual area while trying to start a new research program that, absent a track record of success, might not appeal to proposal reviewers (whether from the pioneer's initial or newfound field) who have the more traditional perspectives of established disciplines. The proposed award would provide start-up research funding for up to 5 years. It would allow cross-disciplinary pioneers the opportunity to engage themselves fully in chemical engineering research so that they can, by the end of the award, contribute to the intellectual life and teaching of their department

and compete with other researchers on an equitable basis in the peer-review system.

Another type of individual award that belongs in the balanced portfolio is the type of "research excellence award" exemplified in the industrial world by the IBM Fellows program. These awards would choose extremely meritorious investigators for research support on the basis of the creativity, productivity, and significance of their recent work, rather than on the basis of a project proposal. They would be provided funding for themselves and one or two associates to pursue topics entirely of their own choosing for a few years. The result is that a select group of the very best researchers would receive the opportunity to explore more speculative and high-potential research ideas. An experimental program of this type, with no more than 15 such awards active at any given time, might leaven the entire field of chemical engineering as well as provide an award that would represent the pinnacle of individual personal achievement in chemical engineering research.

Like other fields where the small research team is still a vital and appropriate funding mode (e.g., chemistry, solid-state physics, biology, and materials science), the cost of conducting cutting-edge research in chemical engineering is growing rapidly. The complex problems of the future will require groups that are larger than today's. The cost of state-of-the-art instrumentation and computational facilities will continue to grow. It is important that agencies funding chemical engineering research ensure that their investment is made most effective by realistically estimating the research costs of groups at the frontiers of the field. In Appendix A (Table A.2), the committee provides its own estimates of appropriate levels of support for groups of different sizes. *Particularly in a time of budgetary stringency, research effectiveness is maximized by funding a smaller number of excellent projects adequately, rather than funding a large number of projects inadequately.*

Funding agencies with vital programs of support for small research groups may also want to consider implementing one of the following two mechanisms:

- Improvement of links between chemical engineering departments and small process technology firms. There is an important problem in some key emerging technologies that is not addressed by existing programs in funding agencies. It deals with the generation and transfer of expertise and ideas from the research laboratory to the production line in biotechnology and process technologies for electronic, photonic, and recording materials and devices. In these areas, a key role in generating new process concepts and equipment is played by a large number of relatively small firms. These firms are capital-poor but rich in problems that would benefit markedly from the insights of academic chemical engineers. The United States could significantly boost its competitive position in these areas by facilitating information transfer between academia and this segment of industry. The problem for funding agencies with an interest in promoting U.S. capabilities in this area is how to create incentives for academic and industrial researchers to seek out and forge links with one another.

Since the partnerships that the committee would like to see fostered will be individualized to the particular needs and interests of the participants, it recommends that any program allow for a wide range of proposed activities. Examples of the sort of initiatives that might be funded include the following: (1) Grants to provide special instrumentation and facilities to be shared by researchers from chemical engineering departments and high-technology firms. The facilities or instrumentation would be located at a university, but available for use by researchers at the high-technology firms. (2) Sabbatical awards for academics at smaller firms that are not likely to be plugged into a large university-based center. (3) Starter grants for industrial liaison programs in chemical engineering departments. A key feature of any of these programs would have to be significant, ongoing person-to-person contact between academic and industrial research groups.

- Large research centers and consortia. Centers are a common mechanism of support to stimulate cross-disciplinary interactions among researchers, or to facilitate cooperation among researchers from academic, industrial, and government laboratories. The most ambitious program of support for cross-disciplinary centers is the NSF Engineering Research Centers (ERC) Program. The cross-disciplinary character of many of the frontiers discussed in this report makes them fruitful areas for center-based research, a reality that has already been recognized by the establishment of ERCs in bioprocess engineering, compound semiconductors, composite materials, hazardous waste management, and computer-assisted design. Chemical engineering plays an important role in nearly all these centers.

Federal agencies should try to stimulate industrial interest in consortia to bolster U.S. industrial competitiveness in technology. DOE should foster interest in consortia to address macroscale energy and natural resource processing research. Here, the experiments that need to be carried out are large in size and costly. Generic research could be carried out in such a way as to (1) allow for fruitful interchange of ideas between industrial and academic researchers, and (2) improve the research efficiency of the energy processing industries while leaving each individual energy company free to develop its own proprietary processes and technologies.

Another area where the large-scale approach might be appropriate is in research related to environmental protection and process safety. Again, this is an area where cooperation on generic research among companies, as well as between industry and academia, focused on generic research problems, could lead to significant research advances. EPA should stimulate such arrangements to expand the amount of research needed in this area.

Recommendations for Specific Federal Agencies

National Science Foundation

The NSF has a logical role in each of the following frontier research areas.

- Biotechnology. NSF should sustain the growth and quality of its existing research support. This area should be targeted by NSF for new support to cross-disciplinary pioneers, partnerships, instrumentation and facilities, and incentives to improve links between academia and small high-technology firms.

• Electronic, photonic, and recording materials and devices. The committee recommends a 5-year pattern of budgetary growth to achieve 30 groups funded at an aggregate level of $6 million per year. This area should also be targeted for new support to cross-disciplinary pioneers and improved links between academia and small high-technology firms.

• Polymers, ceramics, and composite materials. In addition to continued growth in support for research on polymers and polymeric composites, a new thrust is recommended to establish six to eight centers for the chemical engineering of ceramic materials and composites over the next 5 years, funded at a total annual level of $4 million per year. Cross-disciplinary pioneers should also be supported in this area.

• Energy and natural resources processing. NSF should sustain its support of basic research in complex behavior in multiphase systems, catalysis, separations, dynamics of solids transport and handling, and new scale-up and design methodologies.

• Environmental protection, process safety, and hazardous waste management. NSF should strongly support continued growth in this area, focusing on engineering design methodology for process safety and waste minimization. In FY 1986, only three chemical engineering groups working in this area were funded by NSF, with total support of less than $250,000.

• Computer-assisted process and control engineering. The committee recommends a 5-year pattern of growth to a program supporting 35 groups with access to state-of-the-art workstations, software, and computer networks, in addition to the existing ERCs related to this topic. These smaller groups should receive total NSF support of $8 million per year as a target, keyed to additional industry support.

• Surface and interfacial engineering. NSF should expand its support in this area, with emphasis on acquisition by chemical engineers of state-of-the-art instrumentation for surface

and interface studies. The need for such dedicated instrumentation can be met at a funding level of about $5 million per year.

To make its most effective contribution to these areas, NSF should consider the following new support mechanisms for chemical engineers.

• Cross-disciplinary pioneer awards. The committee recommends that a steady-state program of 25 awards in the three areas identified above be achieved over 5 years, at an average level of $100,000 per award per year.

• Research excellence awards. The committee recommends that a steady-state program of 15 awards be achieved over 3 years, at an average level of $100,000 per year per award.

• Improvment of links between chemical engineering departments and high-technology firms. NSF should create a program to provide incentives for partnerships between academic chemical engineers and researchers in small process technology firms in biotechnology and electronics.

Department of Energy

The committee proposes major new initiatives for the Office of Fossil Energy and the Office of Energy Research (OER) to support in-situ processing of resources and the development of tomorrow's liquid fuels for transportation. New initiatives are also proposed for these two offices for the chemical engineering of advanced materials and the development of computational methods and process control—the Division of Engineering and Geosciences in OER should continue and expand its support of fundamental research while the Office of Fossil Energy starts a new program on applications to design and scale-up of large-scale technologies. An initiative is recommended for the Office of Energy Utilization in chemical and process engineering for biotechnology applied to renewable sources for chemical feedstocks. Surface and interfacial engineering should be a strong focus for initiatives in the OER (e.g., catalysis and colloid science and technology) and in Office of Energy Utilization (e.g., corrosion science and electrochemical engineering).

National Institutes of Health

The committee recommends that the National Institutes of Health undertake an initiative to support basic research in chemical and process engineering science for ultimate application to biotechnology and biomedical devices. Such an initiative should be targeted towards (1) encouraging the submission of proposals from cross-disciplinary partnerships between chemical engineers and life scientists and (2) shaping the biochemical engineers of the future by using Institutional National Research Service Awards to facilitate the expansion of graduate course requirements in biochemical engineering to include greater exposure to the life sciences.

Department of Defense

For the Department of Defense (DOD), the main relevant thrust of this report is in the area of materials. The committee recommends that DOD formulate integrated initiatives (i.e., from basic research to testing and evaluation) on problems where improvements in chemical processing is the key to enhanced performance (e.g., polymer-based optical fiber and components; processing for high-strength, high-modulus fibers; manufacturing process technology for composite materials; and joining and repair science for complex materials systems based on polymers, ceramics, and composites). At the basic end of the research spectrum, these would translate into major initiatives for support of molecular science and engineering.

Environmental Protection Agency

The Environmental Protection Agency should revitalize its research grant program in its Office of Exploratory Research and substantially increase its support to chemical engineers investigating important challenges to environmental quality. Stability in the EPA research program over several years is needed to attract the best scientific and engineering research talents to these problems and to allow them to work efficiently on their solution.

The EPA should also consider creating a national "Center for Engineering Research on Environmental Protection and Process Safety" that would provide both unique state-of-the-art laboratory facilities and computational resources to chemical and process engineering researchers from academia, federal laboratories, and industry.

National Bureau of Standards

The small NBS program in chemical engineering should receive substantially greater funding to fulfill critical needs for evaluated data and predictive models. The committee supports NBS plans to focus on data needs in emerging technology areas such as biotechnology and advanced materials.

Bureau of Mines

The committee recommends that the Bureau fund a modest number of university-based centers focused on in-situ processing of dilute resources. This initiative would complement the one proposed for DOE.

RECOMMENDATIONS FOR PROFESSIONAL SOCIETIES

American Institute of Chemical Engineers

Every constituent part of the American Institute of Chemical Engineers (AIChE)—journals, committees, local sections, divisions, and student chapters—should take up the challenge of examining the frontiers presented in this report and, in the context of their mission and purpose, seek out ways of rejuvenating the profession of chemical engineering. The AIChE should set the goal of attracting to its meetings and journals significant numbers of researchers from other disciplines who are working on problems closely related to those at the cutting edge of chemical engineering. The AIChE should provide awards that recognize the achievements of chemical engineering researchers in frontier areas. The Institute should continue to promote reforms in chemical engineering education that would ensure that students would be prepared for new challenges. Finally, the AIChE should undertake cooperative efforts with other organizations to advance the interests of the discipline.

American Chemical Society

Over 12,000 chemical engineers are members of the American Chemical Society (ACS), and ACS meetings, journals, and abstracting services support chemical engineering research in important ways. The findings and recommendations of this report should be of interest and concern to the ACS. In fact, many of the recommendations for the AIChE could be profitably implemented by the ACS Divisions of Industrial and Engineering Chemistry, Colloid and Surface Chemistry, Environmental Chemistry, Fuel Chemistry, Microbial and Biochemical Technology, Polymer Chemistry, and Polymeric Materials: Science and Technology. The ACS is a natural partner for the AIChE in joint undertakings to benefit chemical engineering, and the committee recommends that both organizations pursue opportunities for future joint undertakings.

Council for Chemical Research

The Council for Chemical Research is not a professional society, but it deserves mention in this report because it provides a unique forum for interactions between chemists and chemical engineers and between academic and industrial researchers. It should be credited for building effective bridges among these groups and highlighting the complementary character of chemistry and chemical engineering. The committee recommends that the Council continue to promote these interactions, focusing on the frontiers in this report that are of mutual interest to chemists and chemical engineers.

CONCLUSION

Chemical engineering is a discipline that integrates the research advances of a number of scientific areas. Paramount among these is chemistry, but fields such as applied mathematics, biology, computer science, condensed-matter physics, environmental science and engineering, and materials science also provide important insights that chemical engineers use. This report has highlighted chemical engineering accomplishments and frontiers in a number of areas touched on by these disciplines. It is important that these other disciplines remain vital, too.

Two years ago, the National Research Council published a report on research frontiers and needs in chemistry entitled *Opportunities in Chemistry*, and known colloquially as the "Pimentel Report" after its chairman, George C. Pimentel. That report identified many of the same areas discussed in this report and focused on how chemists contribute to them. This committee endorses the recommendations contained in *Opportunities in Chemistry*, and urges their implementation in addition to the recommendations contained in this volume. The two reports, like the two disciplines, should be seen as complementary, not competing. A vital base of chemical science is needed to stimulate future progress in chemical engineering, just as a vital base in chemical engineering is needed to capitalize on advances in chemistry.

NOTE

1. More detailed suggestions for education and training in separations are contained in *Separations and Purification: Critical Needs and Opportunities* Washington, D.C.: National Academy Press, 1987.

Detailed Recommendations for Funding

CURRENT FUNDING PATTERNS

More than 18 separate federal agencies funded chemical engineering research in FY 1985, the most recent fiscal year for which actual expenditure data are available for all government agencies (Table A.1). Their support to all performers of chemical engineering research—academic, private, and federal—exceeded $254 million. Much of this funding, though, was for research in federal and private laboratories.

Nearly 90 percent of federal support for academic basic and applied research came from only two agencies: NSF and DOE. This narrow funding base is not good for the health of academic research, which is best served by pluralism among funding sources. Neither is it in the best interest of applied and developmental federal programs that depend on chemical engineering. As was observed at a recent conference on research and national priorities:

> A good development manager inevitably runs into fundamental problems requiring a research solution. The road to a solution is easier if the manager has close ties to the research community. The way to maintain such ties is through the maintenance of an ongoing basic research program in fields underlying the development activity.[1]

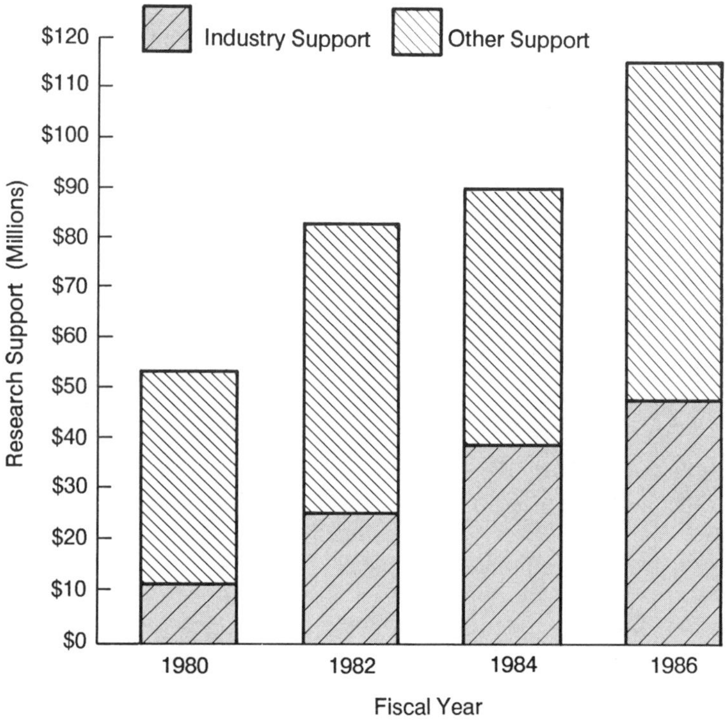

This principle underlies industry's support of academic chemical engineering research. In the last 6 years, industrial support has nearly quadrupled; it has been the main engine for funding growth in academic chemical engineering (Figure A.1). Chemical engineering now leads all engineering disciplines in the proportion of academic support coming from industry and other nonfederal sources (Figure A.2).[2] Industry is often stereotyped as

FIGURE A.1 Industrial support of academic chemical engineering nearly quadrupled from 1980 to 1986 and is the major factor behind growth in academic funding in this period. Data from Council for Chemical Research.

TABLE A.1   Federal Support for Basic and Applied Chemical Engineering Research in FY 1985 (thousands of dollars)

Sponsoring Agency and Subdivision	Support to All Performers	Support to Academia
Department of Agriculture		
Agricultural Research Service	3,267	0
Cooperative State Research Service	2,353	2,353[a]
Department of Defense		
Department of the Army	26,689	1,072
Department of the Navy	24,790	414
Department of the Air Force	1,673	332
Defense Agencies	126	0
Department of Commerce (NBS)	1,660	250[b]
Department of Energy	136,625	15,182
Department of Health and Human Services (NIH)	c	c
Department of the Interior		
Bureau of Mines	c	c
Geological Survey[d]	4,264	0
Minerals Management Service	500	0
Department of Transportation		
Federal Highway Administration	200	0
Federal Railroad Administration	341	0
Research and Special Programs Administration	60	0
Environmental Protection Agency	18,026	520[b]
Federal Emergency Management Agency	200	N/A
National Aeronautics and Space Administration	674	674
National Science Foundation	32,606	27,957
Arms Control and Disarmament Agency	80	N/A
TOTAL	254,134[a]	48,754[a]

[a] Estimate.

[b] Academic data includes development as well as research.

[c] Data for chemical engineering not broken out from data for all engineering research or "engineering, not elsewhere classified."

[d] Includes the Office of Water Research and Technology.

SOURCE: National Science Foundation.[3,4]

being more oriented toward applications and less interested in basic research than federal agencies. Yet the chemical processing industries have invested in academia in a very enlightened manner, even during a major recession in the early 1980s and several quarters of reduced profits or slow growth. Presumably, they are supporting basic research because they believe it will yield insights essential to the long-term profitability of their businesses.

Unfortunately, federal support of academic chemical engineering has grown at only meager rates in the last 6 years. Thus, among engineering disciplines, chemical engineering has simultaneously experienced the second highest rate of growth in industrial support and the lowest rate of growth in federal support (Figure A.3). This is a puzzling pattern to encounter at a time when government support for basic research is widely seen as a way of promoting the international competitiveness of U.S. industry.

Balancing a portfolio of funding sources is always a good idea, but it assumes even more importance as chemical engineering departments seek to expand into new areas. This appendix outlines initiatives for growth in federal programs to respond to the opportunities facing chemical engineers. The initiatives attempt to match research priorities with the missions and purposes of each agency.

TABLE A.2 Costs of Doing Frontier Research in Chemical Engineering

	Annual Level of Effort[a]								NSF[b]
	A		B		C		D		
Faculty summer salary (2 mo), $4,500/mo	2	18.0	1	9.0	1	9.0	1	9.0	8.6
Postdoctoral, $27,000/yr	1	27.0	1	27.0	0.5	13.5	0.5	13.5	1.2
Graduate students, $11,000/yr	8	88.0	6	66.0	4	44.0	2	22.0	12.6
Supplies, $6,000/yr/FTE	(9)	54.0	(7)	42.0	(5)	30.0	(3)	18.0	8.5
Services or other personnel, $3,500/yr/FTE	(9)	31.5	(7)	24.5	(5)	17.5	(3)	10.5	5.1
Equipment and small instruments		10.0		8.0		6.0		5.0	4.9
Indirect costs (34%)		77.7		60.0		40.0		26.5	13.9
TOTAL		306.2		236.5		160.7		104.5	54.8

[a] Under each level of effort are two columns. The first gives multipliers for each budget category, and the second gives subtotals and totals in thousands of dollars.

[b] This column shows the breakdown of costs for the average grant in FY 1986 from the NSF Division of Chemical, Biochemical, and Thermal Engineering (CBTE). The 506 grants made by CBTE supported 508 senior investigators for a total of 95 man-years (2.25 months per investigator) at a cost of $4,340,340; 43 postdoctorals (0.085 per grant) for a total of $622,322 of support; 722 graduate students (1.53 per grant) for a total of $6,396,492 of support; $2,567,871 of other personnel costs; $2,451,796 of equipment and instrumentation; and $4,294,378 of other direct costs. The average indirect cost rate for the CBTE grants was 34 percent.

SOURCE: NSF Directorate of Engineering.

THE COST OF FRONTIER RESEARCH

Before the detailed agency recommendations are presented, some general comments will be made about future research costs.

Pursuing the frontiers described in this report will require more resources than have been needed in the past. Problems of greater complexity require larger research groups to make optimal progress. Access to more sophisticated instrumentation and facilities is costly. Table A.2 presents a range of grant sizes that will be required to perform frontier research efficiently in the future. The estimates in this table for stipends and salaries are reasonable estimates of current costs in chemical engineering departments at major universities. Estimates in the table for costs of services, supplies, and ordinary equipment are close to current NSF averages in chemical engineering.

Four different levels of effort are shown, along with the level of effort supported by the current average grant from the NSF Division of Chemical, Biochemical, and Thermal Engineering (CBTE). Level A shows the costs of a substantial cross-disciplinary partnership be-

tween two research groups. Levels B and C show the costs of large and moderate-sized research groups led by single principal investigators. A group that wants to tackle important research problems at the frontiers of the discipline will probably need to be about the size of Level D to maintain both vitality and continuity.

Throughout the following sections, the cost figures in Table A.2 will be used to derive target budgets for proposed initiatives. It should be stressed, though, that the levels of effort shown in Table A.2 are not meant to imply that there are only four ways to organize a research effort, or to suggest that an explicitly multitiered system of support should be introduced. These levels are meant only to be illustrative of the resources needed today to conduct state-of-the-art research in chemical engineering.

NATIONAL SCIENCE FOUNDATION

The National Science Foundation is the largest source of federal support for academic chemical engineering research (see Table A.1). Its support of the discipline comes through a variety of programs and divisions (Table A.3).

These programs have a logical role in each of the priority areas spelled out in this report, and increased support from NSF is vital to the goal of expanding chemical engineering into new areas. The committee proposes the following new initiatives for the Foundation.

Biotechnology and Biomedicine

Since publication of a preliminary report by this committee in 1984,[5] NSF has increased its support of biochemical engineering by putting in place a new program in biotechnology in its Division of Emerging and Critical Engineering Systems, by funding an Engineering Research Center focused on biotechnology processing, and by increasing support to the CBTE program on biochemical and biomass engineering. The committee applauds this progress and strongly encourages NSF to sustain the growth and quality of its research support in this area.

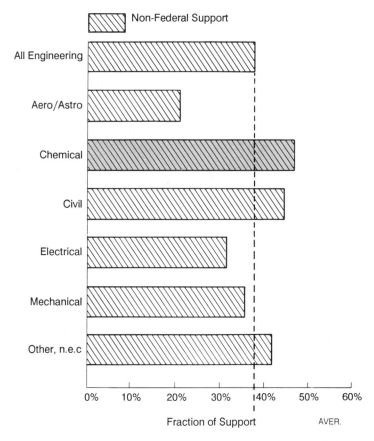

FIGURE A.2 Chemical engineering currently leads all engineering disciplines in the fraction of its support coming from nonfederal sources. Data from National Science Foundation.[2]

NSF Initiative 1

The committee recommends that NSF include biochemical and biomedical engineering in a larger program of 5–year cross-disciplinary pioneer awards (see Chapter 10 for a description). The overall program would be open to candidates from any other discipline with an interest in chemical engineering, and a steady-state program of 25 awards would be achieved over 5 years. The committee recommends that such awards be funded at least at Level D (see Table A.2) with the opportunity to grow to Level C if warranted. This would mean an initial award in the range of $100,000, and a program total of at least $2.5 million at steady state. All candidates for cross-disciplinary pioneer awards would compete in one pool, regardless of their area of interest within chemical engineering. It is likely, however, that some of the awards would be

made to researchers with biological backgrounds who are interested in chemical engineering.

Materials

For materials-related priority areas, the committee recommends small groups or cooperative efforts among small groups as the preferred mode of research organization. There are several reasons for this:

● Many of the frontier research questions outlined in Chapters 4 and 5 can be profitably attacked by adequately supported groups led by a single principal investigator or by multidisciplinary collaborations between small research groups.

● Some important types of research questions

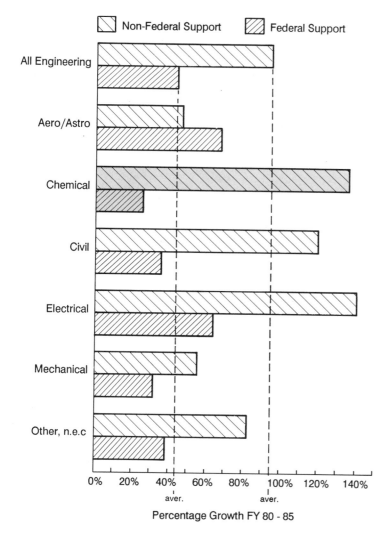

FIGURE A.3 Paradox. Among engineering disciplines, chemical engineering enjoyed the second largest percentage growth in nonfederal support from FY 1980 to FY 1985. During this same period, it also experienced the *lowest* percentage growth in *federal* support. Data from National Science Foundation.[2]

engineering community that the MRLs are more directed toward physics and perhaps not open to significant participation by chemical engineers. It may be less expensive for NSF to investigate the reasons for this perception and to facilitate access by chemical engineers to existing facilities at their home institutions than to create new centers.

● There already is a significant demand in the materials and electronics industry for chemical engineers. These personnel needs are likely to grow as future international competition focuses on materials processing. The existing and anticipated demand for materials-oriented chemical engineers will be most effectively met by a broad-based pattern of support, rather than one concentrated in a few large centers.

NSF Initiative 2

In FY 1986, there were only 15 NSF-supported chemical engineering groups working on the problems of electronic, photonic, and recording materials and devices. Thirteen of these were supported by the Directorate of Engineering and shared a total budget of $755,152. The other two were in the Division of Materials Research of the Directorate of Mathematical and Physical Sciences. Their budgets totaled $211,200. Six of these 15 groups are led by Presidential Young Investigators. *A 5-year initiative should be put in place to double the number of groups working in this critical area.* This is an achievable goal if the best existing groups are allowed to expand in size to produce more faculty candidates, if existing chemical engineering researchers with interests in this area are given the resources to shift their programs, and if some researchers from related disciplines elect to become cross-

require expensive instrumentation and equipment that must be modified extensively by the research team in order to perform its experiments. In such cases, sharing even the same type of equipment among groups with different experimental objectives becomes impossible.

● Creating small groups or collaborations among groups at institutions that have Materials Research Laboratories (MRLs) can be a cost-effective way to add chemical engineering expertise and insights to existing NSF-supported efforts. There is a perception in the chemical

TABLE **A.3** NSF Support of Chemical Engineering in FY 1986
(thousands of dollars)[a]

Directorate, Division, and Program	Total
Directorate of Engineering	
Office of the Assistant Director	247
Electrical, Communications, and Systems Engineering	54
Chemical, Biochemical, and Thermal Engineering	
Kinetics and Catalysis	2,841
Biochemical and Biomass Engineering	2,731
Process and Reaction Engineering	3,248
Multiphase and Interfacial Phenomena	2,156
Separation and Purification Processes	2,870
Thermodynamics and Transport Phenomena	4,322
Thermal Systems and Engineering	2,381
Mechanics, Structures, and Materials Engineering	1,125
Design, Manufacturing, and Computer Engineering	377
Emerging and Critical Engineering Systems	
Biotechnology	2,667
Bioengineering	15
Cross-Disciplinary Research	
Engineering Research Centers	3,638
Industry-University Cooperative Research Projects	243
Industry-University Cooperative Research Centers	508
Directorate of Mathematical and Physical Sciences	
Chemistry	425
Materials Research	3,133
TOTAL	32,981

[a] NSF support of chemical engineering research by all performers.

SOURCE: NSF Directorate of Engineering and NSF Directorate of Mathematical and Physical Sciences.

disciplinary pioneers in chemical engineering departments. An appropriate steady-state group size for research in this area would be somewhere between Levels B and C. This might result in an eventual production of about 40 Ph.D. researchers per year with expertise in the broad range of materials and devices for information storage and handling.

It is somewhat surprising that the average size of the 13 Engineering Directorate grants in this area is only about $58,000. The PYIs are obviously getting industry co-funding, but the total size of individual programs in this area must still be below the optimum level. A minimum target for research support, apart from special instrumentation and facilities, should be about $6 million by the end of the proposed initiative. Industrial co-funding could be required to obtain state-of-the-art equipment or to upgrade facilities.

NSF Initiative 3

The chemical engineering of polymers or composites was the subject of at least 34 grants in the Directorate of Engineering in FY 1986 (totaling $2.37 million), 12 in the Division of Materials Research of MPS (totaling $2.83 million), and an Engineering Research Center grant of $1.25 million. In contrast, there was virtually no identifiable NSF support in FY 1986 for the chemical engineering of ceramics. *In addition to continued growth in support for research on polymers and polymeric composites, a major new thrust is recommended in the chemical engineering of ceramics.* An initial thrust might be to solicit proposals to establish a number of university-based centers on the chemical engineering of ceramics that could then lay the foundation for a more broadly based research effort. Cross-disciplinary interaction between chemists, chemical engineers, and ceramists would have to be a key feature of these centers. One could imagine such centers being about twice the size of a Level A research group. If six to eight such centers were set up over the next 5 years, their steady-state cost (less special equipment and instruments) would be in the range of $4 million per year. In addition to centers, cross-disciplinary pioneers should be supported in this area.

Processing of Energy and Natural Resources

NSF Initiative 4

National Science Foundation programs in catalysis; multiphase systems; separations; dynamics of solids transport and handling; and methodologies for design, scale-up, and control play a key role in supporting more applied research on processing of energy and natural resources. These NSF programs must be sustained and nurtured.

A recent report from the National Research Council recommends that the NSF Separation and Purification Processes Program receive a substantial increment in its budget over the next 5 years.[6] The committee endorses those recommendations.

Environmental Protection, Process Safety, and Hazardous Waste Management

NSF Initiative 5

The National Science Foundation should strongly support growth in this research area, with a special focus on engineering design and control methodology for process safety and waste minimization. In FY 1986, only three chemical engineering groups working in this area were supported by NSF, with combined support of less than $240,000.

Computer-Assisted Process and Control Engineering

NSF Initiative 6

Chemical process systems is one focus of an Engineering Research Center at the University of Maryland established in 1985. In FY 1986, NSF support of the chemical engineering research at this center was reported to be $2.24 million. However, this is the total support received by that center in 1986, and only about 25 percent of the work being carried out is in chemical engineering. In 1986, an Engineering Research Center on design was established at Carnegie Mellon University with an initial grant from NSF of $2.0 million. Again, about 25 percent of the center's effort is in chemical engineering. Thus, NSF has committed to annually fund about $1 million in chemical engineering research in design methodology over the next few years through these two ERCs.

Twenty-two other research groups in computer-assisted process and control engineering, six of which are led by PYIs, received $1.91 million in funding from NSF in FY 1986. While the average grant size for this group of investigators (about $86,800) is much larger than the average grant size for the CBTE Division, a comparison with the levels of effort in Table A.2 shows that in absolute terms these grants still do not provide for a very substantial program. *The committee recommends a major initiative for NSF: a 5-year pattern of growth from the current 22 Level D grants to 35 Level B grants.* These groups will also need access to state-of-the-art workstations, software, and computer networks. Strong co-funding from industry in addition to the NSF Level B grants will help to meet this need, as well as the need for periodic upgrades. At the end of 5 years, the NSF investment in this area, exclusive of the ERCs, should be at a total level of about $8 million. At steady state, this initiative will produce about 50 new Ph.D.s per year with expertise in computer-assisted design, control, and operations. They will have an immense impact on chemical engineering education and practice.

Surface and Interfacial Engineering

NSF Initiative 7

The National Science Foundation should expand its support to surface and interfacial engineering, focusing on surface chemistry, catalysis, electrochemistry, colloid and interfacial phenomena, and plasma chemistry. State-of-the-art research in these areas is very costly, because experimental apparatus must be tailored to individual experiments. Expensive instruments ($200,000 to $500,000) are often extensively modified in the course of studies and become, for all purposes, instruments dedicated to a particular group. For such studies, there are few financial savings to be realized from assembling investigators into centers. *The committee recommends that the NSF provide funds for chemical engineering groups to acquire sophisticated instrumentation for studying surfaces, interfaces, and microstructures.*

The committee estimates that, in a given year, somewhere between 10 and 25 of the active groups in surface and interfacial engineering will need to acquire a major instrument for adaptation and use. A funding level of $5 million per year for major dedicated instrumentation can meet most of these needs.

Research Excellence Awards

NSF Initiative 8

The committee recommends that a steady-state program of 15 research excellence awards in chemical engineering be achieved over 3 years. This new mechanism is described in Chapter 10. Because these awards are intended to fund speculative high-potential research, they will probably work best in the milieu of a small research group. Thus, the committee recommends that they be funded at about Level D. The steady-state cost of the initiative would be $1.5 million.

Conclusion

The committee's recommendations for NSF target about $22 million in growth over the next 5 years in six major initiatives, and a less determinate amount of growth in the other two initiatives. The six major initiatives would add 84 new research groups to chemical engineering over 5 years, a 17 percent increase in the number of groups funded. In terms of dollars, the initiatives would amount to a rough doubling of the amount that chemical engineering research received from the CBTE Division in FY 1986.

DEPARTMENT OF ENERGY

The Department of Energy has wide-ranging programmatic interests to which chemical engineering can make important contributions. These include familiar areas of fossil resource production and processing, catalysis, separations, and nuclear energy. They also include less familiar areas such as materials, process design and control, and molecular biology.

In-Situ Processing of Energy and Mineral Resources

DOE Initiative 1

The committee's prime initiative for DOE is the support of research on fundamental phenomena important for in-situ processing of hydrocarbon resources. (A related initiative for the Bureau of Mines is discussed later in this appendix.) The important fundamental problems in this area are outlined in Chapter 6. The size and extreme complexity of the environments in which these phenomena occur will require expensive, large-scale, prolonged field experiments. Such large-scale research, though, will be quite different from the demonstration projects funded by DOE in the late 1970s and early 1980s. Rather than demonstrating the maturity of technologies and their readiness for commercial application (an activity in which, it has been argued, the federal government should not be involved), the focus of large-scale fundamental research proposed for this initiative would be to build a nonproprietary knowledge base relating experimental results on the smaller scales of test systems and equipment to results obtained in the larger and more complex systems found in the field.

A variety of support mechanisms for carrying out such research could be envisioned that would include sponsorship of individual research projects in academia or federal laboratories, where appropriate; a DOE equivalent of the NSF Engineering Research Centers, but with more cooperative involvement from industry; and stimulation by DOE of industrial consortia both to carry out joint research among companies on nonproprietary topics and to support relevant research in academia. The importance of this research is such that substantial interest in cooperative research might be generated in industry if DOE took a lead role in providing stimulus and partial funding.

Liquid Fuels for the Future

DOE Initiative 2

A second research initiative for DOE centers on facilitating progress towards the next generation of liquid fuels. This is a very broad topic, encompassing many areas including catalysis (see Chapter 9), solids processing, separations, materials development, and advanced scale-up and design techniques. The Office of Energy Research (OER) and the Office of Fossil Energy should work together to coordinate research in these areas. Some research areas, notably solids processing, may require the same type of large-scale fundamental research called for in the previous research initiative. The mechanisms proposed there for stimulating large-scale research may be applicable here, as well.

Advanced Computational Methods and Process Control

DOE Initiative 3

The Division of Engineering and Geosciences in OER supports cutting-edge research in fluid dynamics and process design and control. These programs should be sustained as a vital part of the balanced portfolio of support for these areas within chemical engineering. Process design, scale-up, and control have already been mentioned as important keys to in-situ processing. The Office of Fossil Energy should consider setting up a research program in this area that would support fundamental process design and control research that would be particularly applicable to large-scale projects.

Surface and Interfacial Engineering

DOE Initiative 4

The Division of Chemical Sciences in OER supports basic chemical research. The primary involvement of chemical engineers with this

program has been in the areas of catalysis and separations. Given the broad range of energy applications in which the structure and chemistry of interfaces is important, the committee recommends that the Division undertake an initiative in the chemical control of surfaces, interfaces, and microstructures. This would include support of work by both chemists and chemical engineers in the areas of surface chemistry, plasma chemistry, and colloid and interfacial chemistry.

Microstructured Materials

DOE Initiative 5

Materials in general, and ceramics in particular, are heavily emphasized in the DOE Division of Material Sciences. Up to now, this program has had relatively little involvement from chemical engineers. Given that chemical processing approaches to ceramics have a bright future, both for structural applications and possibly for ceramic superconductors, the OER should consider a major thrust in chemical processing of materials, with a view towards the more facile production of defect-free ceramics for energy and energy-saving applications.

Biotechnology

DOE Initiative 6

Bioprocessing is of interest to the DOE Division of Energy Conversion and Utilization Technologies (ECUT), which is concerned with increasing the efficiency of energy conversion and the use of renewable resources. A recent report of the National Research Council proposes a comprehensive program in this area for ECUT, involving both chemical engineering and the life sciences, and funded for 10 years at an annual level of $10 million.[7] The committee endorses these recommendations and urges their implementation.

NATIONAL INSTITUTES OF HEALTH

The National Institutes of Health is the premier sponsor of health-related research in the

United States. Its long-term support of the basic biosciences is responsible for the advances that have made biotechnology possible. Sophisticated engineering will be needed, though, if biotechnology is to make its full potential contribution to the nation's health. NIH supports a great deal of research aimed at elucidating molecular processes in living systems; identifying molecules of potential therapeutic value; and developing potential routes to them, whether via synthetic chemistry or recombinant DNA organisms. It provides less support to the problem of turning these potential synthetic routes into practical, economic processes. In part this is because NIH has traditionally focused on basic science, leaving commercialization of discoveries to others. But there is a knowledge gap in the basic engineering science for biotechnology and biomedicine that is not being filled by industry. This gap impedes the full conversion of new biological knowledge into products and therapies for improving human health. There is a role for NIH, consistent with its historical mission and philosophy, to (1) expand the base of fundamental knowledge in the chemical engineering of biological systems and (2) train a new generation of chemical engineers to be more conversant with the biological sciences. Both steps will allow chemical engineers of the future to expertly apply engineering principles to biological problems.

NIH Initiative 1

The committee recommends that NIH undertake an initiative to bring chemical engineering researchers into more effective contact with biological and medical researchers. The mechanism for accomplishing this would be through cross-disciplinary partnerships in research, with groups ranging upwards in size from a traditional research project led by two co-principal investigators—one from chemical engineering and one from the life sciences—to program projects and perhaps even centers involving many investigators from engineering, biology, and medicine. At the low end of this spectrum, a level of funding similar to Level A in Table A.2 would allow for a substantial interdisciplinary effort.

NIH Initiative 2

The National Institutes of Health could also play a vital role in shaping the biological chemical engineers of the future by using Institutional National Research Service Awards to provide biochemical engineering graduate students with a greater exposure to the life sciences. A significant limiting factor in expanding the graduate curriculum in this way is the problem of supporting students for an additional year prior to their immersion into grant-supported research.

DEPARTMENT OF DEFENSE

The chemical engineering research frontiers of most relevance to the Department of Defense are in materials. Faster electronic devices, more reliable communication systems, and stronger structural components are all needed by DOD in order to fulfill its mission. Chemical processing is a valuable tool to tailor these materials for specific military uses.

A special strength of the DOD research infrastructure is its vertical integration from basic research, through applied and exploratory research, to advanced testing and evaluation of technologies in the field. *The committee recommends that DOD exploit this strength by formulating integrated initiatives around topics where advances in chemical processing can exert leverage.* The development of more secure signal and communication systems might be one such topic. Chemical processing plays an important role in the manufacture of both glass-based and polymer-based optical fibers. The latter are more easily fit to connectors and attached to one another. They might be most appropriate for defense systems requiring transmission of data over short distances (measured in meters) rather than long distances. An initiative to develop practical polymer-based communications systems would require a significant basic research effort in polymer chemistry and chemical engineering to resolve materials challenges to low-loss optical transmission in polymers, a significant effort in chemical engineering to provide the fundamental insights needed to fabricate optical-quality polymer lenses, splitters, and connectors, as well as research in related disciplines (e.g., electrical engineering) to develop overall system concepts and design and to produce a system that could be integrated into existing hardware. The potential payoffs of success in this initiative would be enormous, both in terms of national security (e.g., secure optically based communications systems cheap enough to install nationwide) and in terms of society's need for rapid, efficient transmission of data (e.g., optically based local telephone systems and local area networks capable of simultaneously transmitting enormous quantities of digitized voice and data signals).

This is just one example of several such DOD initiatives that might be built around scenarios that assume that basic chemical problems in materials could be solved by concentration of resources on fundamental research. One can just as easily imagine other initiatives, such as the following:

• The high strength of Kevlar® fibers is due to the way in which they are processed, rather than their intrinsic chemical composition. If processes could be found that were capable of creating in other materials the highly ordered structure seen in Kevlar®, it might be possible to fabricate extremely durable treads for use in modern infantry vehicles and tanks. Such a practical focus might be used as an organizing focus for a substantial program of fundamental research on polymer processing for high strength and toughness.

• Most composites used in aircraft must be "laid up" by hand, because a reliable manufacturing technology for composites has yet to be discovered. Chemical processing combined with textile engineering could be used to achieve major advances in the manufacture of reliable composites for major structural components of aircraft.

• Another composites problem is joining and repairing these materials. Unlike metals, where patches can literally be bolted onto a system without substantially degrading performance, performance in composites is very sensitive to the means by which composite components are joined to each other, or repaired. The practical problem of repairing the composite aircraft of the future could be the focus of a significant fundamental chemical engineering effort to elu-

cidate joining and repair on the molecular level, and to integrate new insights from the molecular level into the contributions that would be made on the systems level by other disciplines (e.g., materials, mechanical, and aerospace engineering).

ENVIRONMENTAL PROTECTION AGENCY

EPA Initiative 1

The Environmental Protection Agency needs a strong basic research program, especially in chemical science and technology. The committee urges EPA to revitalize its research grant program in the Office of Exploratory Research. As part of this revitalization, EPA should seek to fund the best chemical engineering research groups investigating important ongoing challenges to environmental quality:

• fundamental chemical processes important in the generation and control of toxic substances by combustion,

• chemical processes involved in the transport and fate of hazardous substances in the environment, and

• design methodologies that could result in waste and process hazard minimization in chemical manufacturing plants.

These areas all promise substantial advances in improving our environment, but will not yield that promise in an atmosphere of on-again, off-again funding. Stability in the EPA research program over several years is needed to attract the best scientific and engineering research talents to these problems and to allow them to work efficiently on their solution.

EPA Initiative 2

The EPA should consider establishing a national Center for Engineering Research on Environmental Protection and Process Safety (CERES), modeled on the National Center for Atmospheric Research. Chemical and process engineering researchers would benefit from a special collection of state-of-the-art laboratory facilities and computational resources dedicated to research on environmental protection, process safety, and hazardous waste management. As a centralized facility, CERES could serve a coordinating role to enhance cooperation in research across institutional boundaries and to diffuse rapidly into industry research advances made in academic and other laboratories. The specific tasks and possible organization of such a center, as well as its potential relationship with a new NSF center on hazardous waste management, should be the subject of an in-depth study by EPA and any competition for siting and operating this center should be open to the most meritorious proposal, whether it originates from a university, a federal laboratory, the nonprofit sector, or some combination of the three.

NATIONAL BUREAU OF STANDARDS

The National Bureau of Standards has a unique role to play in supporting the field of chemical engineering. It should be the focal point for providing evaluated data and predictive models for data to facilitate the design, the scale-up, and even the selection of chemical processes for specific applications.

Despite the plethora of data in the scientific literature on thermophysical quantities of substances and mixtures, many important data gaps exist. Predictive capabilities have been developed for problems such as vapor-liquid equilibrium properties, gas-phase and—less accurately—liquid-phase diffusivities, and solubilities of nonelectrolytes. Yet there are many areas where improved predictive models would be of great value. An accurate and reliable predictive model can obviate the need for costly, extensive experimental measurements of properties that are critical in chemical manufacturing processes.

Particular attention should be given by NBS to data needs in the emerging technology areas served by chemical engineering (i.e., biotechnology and materials). In the area of biotechnology, the NBS is attempting to identify and assign priority to the thermophysical properties of greatest importance to scale-up and commercialization, and to identify promising theoretical approaches that could lead to generic predictive models for the types of complex mixtures found in bioprocessing systems. The committee endorses this effort and encourages the NBS to provide the needed level of funds for an optimal effort. In the materials area, the need for international standards for advanced materials, such as polymer blends and ceramics, is acute. Again, the NBS has started an effort in this area as part of the international Versailles Project on Advanced Materials and Systems (VAMAS). Currently, about 100 U.S. researchers are involved in VAMAS-related research, in both industry and academia. The amount of federal funding for this effort, though, is less than $500,000. This type of project is extremely important to the rapid worldwide development of advanced materials, and should be funded at a level more commensurate with that importance.

BUREAU OF MINES

The Bureau of Mines, within the Department of the Interior, funds a substantial amount of chemical engineering research in its in-house laboratories, particularly in the area of hydrometallurgical separation processes. The U.S. minerals industry is currently in a depressed state typified by diminished research efforts within industrial laboratories and, in some cases, wholesale termination of research operations. As a result, new researchers have bleak prospects for industrial employment. At the same time, the United States cannot afford to lose a professional generation of research personnel in an area that would be of critical importance if foreign supplies of certain metals were interrupted.

The committee recommends that the Bureau fund a modest number of university-based centers focused on in-situ processing of dilute resources. This initiative would complement the major one proposed for DOE. Such centers should explicitly focus on generic themes, such as separations from highly dilute solutions, multiphase flow though porous media, or the development of sensors and other instrumentation. The goal of the centers program would be to stimulate fresh ideas and insights in metals-related processing research and to train a new generation of research engineers flexible enough either to move into a revitalized minerals industry or to find employment in the broader sector of process industries.

NOTES

1. National Academy of Sciences, Government-University-Industry Research Roundtable. *What Research Strategies Best Serve the National Interest in a Time of Budgetary Stress? Report of a Conference*. Washington, D.C.: National Academy Press, 1986.

2. National Science Foundation, Division of Science Resources Studies. *Academic Science/Engineering: R&D Funds, Fiscal Year 1985*. Washington, D.C.: National Science Foundation, 1986.

3. National Science Foundation, Division of Science Resources Studies. *Federal Funds for Research and Development, Fiscal Years 1985, 1986, and 1987*, Volume XXXV (Detailed Statistical Tables).

4. National Science Foundation, Division of Science Resources Studies. *Federal Support to Universities, Colleges, and Selected Nonprofit Institutions, Fiscal Year 1985*. Washington, D.C.: National Science Foundation, 1987.

5. National Academy of Sciences-National Academy of Engineering-Institute of Medicine, Committee on Science, Engineering, and Public Policy. "Report of the Research Briefing Panel on Chemical and Process Engineering for Biotechnology," in *Research Briefings 1984*. Washington, D.C.: National Academy Press, 1984.

6. National Research Council, Committee on Separation Science and Technology. *Separation and Purification: Critical Needs and Opportunities*. Washington, D.C.: National Academy Press, 1987.

7. National Research Council, National Materials Advisory Board. *Bioprocessing for the Energy-Efficient Production of Chemicals*. Washington, D.C.: National Academy Press, 1986.

Contributors

The Committee on Chemical Engineering Frontiers: Research Needs and Opportunities would like to express gratitude to the following individuals who provided suggestions, presentations, written submissions, or critiques that were used in the preparation of this report. In many cases, these inputs were solicited by one of the panels involved in the writing of this report. The organizational affiliation of contributors at the time they provided their input is also listed. Contributors of figures and photographs for the report are acknowledged where their submissions appear. The committee, of course, is responsible for the final content of the report.

ABRAMOWITZ, P. H., St. Joe Minerals Corporation
AGARWAL, J., Charles River Associates
AHEARNE, J. H., Resources for the Future
ALKIRE, R. C., University of Illinois
ALLRED, V. D., Marathon Oil Company
ALPAN, F. F., Pennsylvania State University
ALPERT, S. B., Electronic Power Research Institute
ANDERSON, H. R., IBM
ANDERSON, T. J., University of Florida
ANGUS, J. C., Case Western Reserve University
APPLEBY, A. J., Electric Power Research Institute
ARKUN, Y., Rensselaer Polytechnic Institute
ASTROM, K. J., Lund Institute of Technology
ATHERTON, R. W., Sentry Computer Integrated Manufacturing Systems
AZIZ, K., Stanford University
BAEDER, D. L., Consultant
BAILEY, J. E., California Institute of Technology
BAXTER, J., Illinois State Geological Survey
BENNION, D. N., Brigham Young University
BLANCH, H. W., University of California, Berkeley

BOHRER, M., AT&T Bell Laboratories
BORTZ, S. A., IIT Research Institute
BOSANQUET, L., Monsanto Company
BOUDART, M., Stanford University
BRATON, R., IBM
BRIAN, P. L. T., Air Products and Chemicals, Inc.
BRISTOL, E. H., Foxboro Company
BROWN, R. A., Massachusetts Institute of Technology
BROWNAWELL, D., Exxon Chemical Company
BUILDER, S. E., Genentech, Inc.
BURGER, R. M., Semiconductor Research Corp.
CHIEN, H. H., Monsanto Company
COOPER, W., Eastman Kodak Company
COX, R. K., E.I. du Pont de Nemours & Co., Inc.
CROWE, C. M., McMaster University
DAHLSTROM, D. A., University of Utah
DEEN, W. E., Massachusetts Institute of Technology
DOHERTY, M. F., University of Massachusetts
DORSCH, R. R., E.I. du Pont de Nemours & Co., Inc.

DOSCHER, T. M., The Doschers Group
DOSS, J. E., Tennessee Eastman Company
ECONOMY, J., IBM Almaden Research Center
EGGERT, R. G., Pennsylvania State University
EIDEL, J., Illinois State Geological Survey
ERLINGER, H., Illinois State Geological Survey
EVANS, J. W., University of California
FAULKNER, L., University of Illinois
FEFFERMAN, G. B., AT&T Bell Laboratories
FISHER, D. G., University of Alberta
GEORGAKIS, C., Lehigh University
GORBATY, M. L., Exxon Research & Engineering Co.
GREEN, D. W., University of Kansas
GREGOLI, A. A., Cities Service Research Company
GUTOFF, E. B., Polaroid Corp.
HALE, J. C., E.I. du Pont de Nemours & Co., Inc.
HAMMOND, G. S., Allied Corporation
HANISCH, W., Cetus Corporation
HASHIMOTO, I., Kyoto University
HELT, J. E., Argonne National Laboratory
HENCH, L. L., University of Florida
HENLEY, E. J., University of Houston-University Park
HENRIE, T. H., U.S. Bureau of Mines
HERBST, J., University of Utah
HIRASKI, G. J., Shell Development Company
HLAVACEK, V., State University of New York, Buffalo
HOPFENBURG, H. L., North Carolina State University
HORWITZ, E. P., Argonne National Laboratory
HOWES, M. A. H., IIT Research Institute
HUGHES, R., Illinois State Geological Survey
HUGHES, T. R., Chevron Research Company
HUNT, A. J., Lawrence Berkeley Laboratory
IRVING, J. P., Chevron Oil Field Research Co.
JACQUES, D., Exxon Chemical Company
JEFCOAT, I. A., University of Alabama
JEZL, J. L., Amoco Chemicals Company
JEROME, J., Northwestern University
JOHNSON, D. W., AT&T Bell Laboratories
JONES, F. N., North Dakota State University
JUBA, M. R., Eastman Kodak Company

KAHN, L., Illinois State Geological Survey
KALELKAR, A. S., Arthur D. Little, Inc.
KALHAMMER, F., Electric Power Research Institute
KARDOS, J. L., Washington University, St. Louis
KELLOG, H. J., Columbia University
KEMP, H. S., E.I. du Pont de Nemours & Co., Inc.
KING, A. G., Ferro Corporation
KING, C. J., University of California
KLEIN, L. C., Rutgers University
KNIGHTS, J., Xerox Palo Alto Research Center
KRAMER, M. A., Massachusetts Institute of Technology
LARRABEE, G. B., Texas Instruments, Inc.
LEES, J. K., E.I. du Pont de Nemours & Co.
LEESLEY, M. E., Austin, Tex.
LEIGHTON, M. W., Illinois State Geological Survey
LEWIS, D., GAF Corporation
LEWIS, I. C., Union Carbide Corporation
LIN, CHENG-YIN, AT&T Bell Laboratories
LIU, H. T., Union Carbide Corporation
LYNCH, R. W., Sandia National Laboratory
MACDONALD, S., IBM Almaden Research Center
MAH, R. S. H., Northwestern University
MALLINSON, J. C., University of California, San Diego
MANZIONE, L. T., AT&T Bell Laboratories
MARGOLIN, S. V., Consultant
MATHIS, J. F., NL Industries
MATTSON, J. F., Sepracor, Inc.
MAZDIYASNI, K. S., Wright-Patterson Air Force Base
MEMERING, M. N., Pfizer Pigments, Inc.
MERRIMAN, J. R., Union Carbide Corporation
MERZ, J. L., University of California, Santa Barbara
MILLER, J., University of Utah
MILLER, S. D., Ampex Corporation
MORARI, M., California Institute of Technology
MORRISON, D. L., IIT Research Institute
MOTARD, R. L., Washington University
MUNTER, J. D.
MURATA, H., The Furukawa Electric Co., Ltd.

NAGEL, S. R., AT&T Bell Laboratories
NEWMAN, J. S., University of California
NISHIMURA, A., O Konica Technology Inc.
O'BRYAN, H. M., JR., AT&T Bell
    Laboratories
OTTO, R. J., Lawrence Berkeley Laboratory
OUBRE, C., Shell Development Company
OVERCASH, M. R., North Carolina State
    University
PALSSON, B., University of Michigan
PEPPAS, N. A., Purdue University
PETRICK, M., Argonne National Laboratory
PHILLIPS, J., Boeing Computer Services, Inc.
PIMENTAL, D., Cornell University
PIPES, R. B., University of Delaware
POSNIK, A., Ferro Corporation
PRATER, D., California Institute of
    Technology
PRETT, D. M., Shell Development Company
PROGELHOF, R. C., AT&T Bell Laboratories
PRUD'HOMME, R. K., Princeton University
RAJAGOPALAN, R., Rensselaer Polytechnic
    Institute
RANKIN, J. D., Imperial Chemical Industries
RAWSON, N. E., IBM
RAY, W. H., University of Wisconsin
REKLAITIS, G. V., Purdue University
RICE, R. W., W. R. Grace and Company
RIPPIN, D., ETH-Zentrum
ROBERTSON, J. L., Exxon Research and
    Engineering Co.
RODRIGUEZ, F., Cornell University
ROGERS, K., National Science Foundation
ROLF, M. J., Owens/Corning Fiberglas
ROSENBERG, R. B., Gas Research Institute
ROSSEN, R. H., Exxon Production Research
    Company
ROSTENBACH, R. E., National Science
    Foundation
RUSHAK, K., Eastman Kodak Co.
SATHER, N. F., Argonne National Laboratory
SAWICKI, E., Microsafe, Inc.
SAXENA, A. N., Gould AMI Semiconductors
SCHOENBERGER, R. J., Roy E. Weston, Inc.
SCHWARTZ, B., IBM
SCOTT, C. D., Oak Ridge National Laboratory

SCOTT, J. W., Chevron Research Company
    (retired)
SHAH, S. L., University of Alberta
SHAH, Y. T., University of Pittsburgh
SHAW, D., Texas Instruments, Inc.
SHAW, H., New Jersey Institute of
    Technology
SIEG, R. P., Chevron Research Company
    (retired)
SKALNY, J. A., W. R. Grace and Company
SMYRL, W. H., University of Minnesota
SMITH, D., E. I. du Pont de Nemours & Co.,
    Inc.
SNEDIKER, D. K., Battelle Columbus
    Laboratories
SOHN, H. Y., University of Utah
SOLOMON, P. R., Advanced Fuel Research
SQUIRES, A. M., Virginia Polytechnic
    Institute
STADTHERR, M. A., University of Illinois
STEINDLER, M. J., Argonne National
    Laboratory
STEPHANOPOULOS, G., Massachusetts
    Institute of Technology
STEVENSON, F. D., U.S. Department of
    Energy
TEBO, P. V., E. I. du Pont de Nemours &
    Co., Inc.
THEMELIS, N. K., Columbia University
THURSTON, C. W., Union Carbide
    Corporation
TOBIAS, C. W., University of California
TRACHTENBERG, I., Texas Instruments, Inc.
ULMER, K., Genex Corporation
UMEDA, T., Chiyoda Chemical Engineering &
    Construction Co., Ltd.
VALENTAS, K., Pillsbury Corporation
WADSWORTH, M., University of Utah
WANG, D. I. C., Massachusetts Institute of
    Technology
WENDT, J. O. L., University of Arizona
WHITE, J. R., Consultant
WHITELEY, R. L., Research and Innovation
    Management
WHITESIDES, G. M., Harvard University
WYMER, R. G., Oak Ridge National
    Laboratory

APPENDIX C

The Chemical Processing Industries

Any discussion of the chemical industry must make clear how that industry is being defined. Chemistry, chemical processing, and chemicals are ubiquitous in modern society and industry. In this report, where the term "chemical industry" is used in connection with statistics or economic data, the data refers to the industry segment defined by Standard Industrial Classification (SIC) code 28. Where the term "chemical processing industries" or "CPI" is used in this report, a broader range of industries is meant. The definition of the CPI adopted by this report is the same one used by Data Resources, Inc., and *Chemical Engineering* magazine in their reporting on industry. The industry segments and SIC codes subsumed by this definition of the CPI have been supplied to the committee by *Chemical Engineering* and are reprinted here with permission.

Industry Segment	SIC Code
Chemicals (including petrochemicals)	
Alkalies and chlorine	2812
Industrial gases	2813
Industrial inorganic chemicals, n.e.c.*	2819
Synthetic rubber (vulcanizable elastomers)	2822
Gum and wood chemicals	2861
Cyclic crudes and intermediates	2865
Industrial organic chemicals, n.e.c.	2869
Chemicals and chemical perparations, n.e.c.	2899
Plastics materials, synthetic resins, and nonvulcanizable elastomers	2821
Man-made fibers	
Cellulosic man-made fibers	2823
Synthetic organic fibers, except cellulosic	2824
Drugs and cosmetics	
Biological products	2831
Medicinal chemicals and botanicals	2833
Pharmaceutical preparations	2834
Perfumes, cosmetics, and other toilet preparations	2844

Industry Segment	SIC Code
Soap, glycerin, cleaning, polishing, and related products	
Soap and other detergents except specialty cleaners	2841
Specialty cleaning, polishing, and sanitation preparations, except soap and	
detergents	2842
Surface active agents, finishing agents, sulfonated oils, and assistants	2843
Paints, varnishes, pigments, and allied products	
Inorganic pigments	2816
Paints, varnishes, lacquers, enamels, and allied products	2851
Fertilizers and agricultural chemicals	
Nitrogenous fertilizers	2873
Phosphatic fertilizers	2874
Fertilizers, mixing only	2875
Pesticides and agricultural chemicals, n.e.c.	2879
Petroleum refining and coal products	
Petroleum refining	2911
Paving mixtures and blocks	2951
Asphalt felts and coatings	2952
Lubricating oils and greases	2992
Coke and by-products: part of SIC 3312—"Blast furnaces (including coke	
ovens), steel works, and rolling mills"	3312
Rubber products	
Tires and inner tubes	3011
Rubber footwear	3021
Reclaimed rubber	3031
Fabricated rubber products, n.e.c.	3069
Leather tanning and finishing	3111
Foods and beverages	
Condensed and evaporated milk	2023
Wet corn milling	2046
Cane sugar, except refining only	2061
Beet sugar	2063
Malt beverages	2082
Malt	2083
Wines, brandy, and brandy spirits	2084
Distilled, rectified, and blended liquors	2085
Flavoring extracts and syrups, n.e.c.	2087
Roasted coffee, includes instant coffee	2095
Food preparations, n.e.c.	2099
Fats and oils	
Cottonseed oil mills	2074
Soybean oil mills	2075
Vegetable oil mills, n.e.c.	2076
Animal and marine fats and oil	2077
Shortening, table oils, margarine, and other edible fats and oils, n.e.c.	2079

Industry Segment	SIC Code
Wood, pulp, paper, and board	
Wood preserving	2491
Pulp mills	2611
Paper mills, except building paper mills	2621
Paperboard mills	2631
Paper coating and glazing	2641
Building paper and building board mills	2661
Stone, clay, glass, and ceramics	
Flat glass	3211
Glass containers	3221
Pressed and blown glass and glassware, n.e.c.	3229
Brick and structural clay tile	3251
Ceramic wall and floor tile	3253
Clay refractories	3255
Structural clay products, n.e.c.	3259
Pottery and related products	3261
	3262,
	3263,
	3264,
	3269
Gypsum products	3275
Abrasive products	3291
Asbestos products	3292
Steam and other packing, and pipe and boiler covering	3293
Minerals and earth, ground or otherwise treated	3295
Mineral wool	3296
Non-clay refractories	3297
Nonmetallic mineral products, n.e.c.	3299
Lime and cement	
Cement, hydraulic	3241
Lime	3274
Metallurgical and metal products	
Electrometallurgical products	3313
Primary smelting and refining of copper	3331
Primary smelting and refining of lead	3332
Primary smelting and refining of zinc	3333
Primary production of aluminum	3334
Primary smelting and refining of nonferrous metals, n.e.c.	3339
Secondary smelting and refining and alloying of nonferrous metals and alloys	3341
Enameled iron and metal sanitary ware	3431
Electroplating, plating, polishing, anodizing, and coloring	3471
Coating, engraving, and allied services, n.e.c.	3479
Explosives and ammunition	
Explosives	2892

Industry Segment	SIC Code
Small arms ammunition	3482
Ammunition, except for small arms, n.e.c.	3483
Ordnance and accessories, n.e.c.	3489
Other chemically processed products	
Part of broad woven fabric mills, wool, including dyeing and finishing	2231
Finishers of broad woven fabrics of cotton	2261
Finishers of broad woven fabrics of man-made fiber and silk	2262
Dyeing and finishing textiles, n.e.c.	2269
Artificial leather, oilcloth, and other impregnated and coated fabrics, except rubberized	2295
Adhesives	2891
Printing ink	2893
Carbon black	2895
Carbon and graphite products	3624
Semiconductors and related devices	3674
Storage batteries	3691
Primary batteries (dry and wet)	3692
Part of photographic equipment and supplies	3861
Lead pencils, crayons, and artists materials	3952
Carbon paper and inked ribbons	3955
Linoleum, asphalted felt base, and other hard surface floor covering, n.e.c.	3996
Manufacturing industries, n.e.c.	3999
Engineering design/construction	
Engineering firms, engineering and construction firms, consulting engineers, independent R&D, central engineering, and others	891

*n.e.c. = not elsewhere classified.

Index

A

Adhesives
 manufacturing problems, 69–70
 research needs on, 166
 U.S. competitiveness in, 71
 uses, 69, 76
Adipic acid from microbial fermentation of glucose, 26
Agriculture
 biologically derived pesticides, 23
 opportunities for chemical engineers in, 17, 22–23
 pollution from, 108
 product markets, 18
 veterinary pharmaceuticals, 23
Air pollution
 indoor, 126
 modeling, 14, 125, 126, 142
 monitoring, 125–126
 reduction strategies, 113–117
Aircraft
 ceramic applications in, 65–66
 composites applications, 62, 64, 71, 76
American Chemical Society, recommended role of, 184
American Institute of Chemical Engineers
 Center for Chemical Process Safety, 175, 178
 continuing education program, 77
 promotion of cross-disciplinary cooperation, 34
 recommended role of, 183–184
Antibiotics, penicillin, 10, 11–12
Antibodies, 20, 22, 27
Antigens, 19, 22
Antihemophilic factors, 22
Artificial organs, tissues, and fluids
 aqueous and vitreous humors, 20
 chemical engineering contributions to development of, 10, 19–20
 future targets, 19–20
 heart program, 10
 hybridization, 19
 kidney program, 19
 pancreas, 20, 169
 performance forecasts, 31
 skin for burn patients, 20
 "smart membrane" device, 20
 surface and interfacial phenomena in, 155
 synovial fluids in joints, 20
 see also Prostheses
Ash from combustion processes, 98, 115, 116–117, 156
Automobiles
 composites applications, 64, 71
 electric vehicle technology, 95
 emissions, 108
 polymer applications, 14

B

Biochemical processes in humans, measurement of, 31
Biochemical synthesis
 opportunities for chemical engineers in, 23–24
 see also Cell/tissue culture
Bioengineering
 curricula, 32
 faculty needs, 33
 funding for research, 26, 32–33, 180
 instrumentation and facility needs, 33–34
 manpower needs, 34
 surface and interfacial phenomena in, 2, 17, 27–28, 155
Biological systems, complex
 engineering analysis of, 2, 30–31
 interactions, modeling, 2, 17, 26–27
Biomedicine
 clinical implants and biomedical devices, 27, 31
 contributions of chemical engineers to, 19–20
 diagnostics, 18–20, 169
 educational/training needs for engineers, 32–33, 176
 international competition in, 17, 25–26
 membrane technology applications, 20, 168–169
 opportunities for chemical engineers in, 2, 17, 18–31
 product market, 18
 research recommendations, 27, 30–31
 therapeutics, 19, 21–23, 30–31, 154, 163
 see also Artificial organs, tissues, and fluids; Prostheses
Bioprocessing
 batch, 20, 28–29
 continuous, 2, 29
 high-fructose corn syrup, 24–25
 monitoring and control, 2, 29; *see also* Sensors
 of monoclonal antibodies, 20

205

M

equipment design and scale-up, 4, 99, 100, 143
  interdisciplinary cooperation, 100
  research needs, 99–100, 103
  scale-up factors, 103
Soot from combustion processes, 114–116
Soviet Union, fission research, 95
Spectroscopic methods
  Auger electron, 158, 162, 172
  carbon-13 NMR, 168
  catalyst characterization, 158, 159, 171
  electron energy loss, 159, 171
  electron spin resonance, 163
  emission, 163
  extended x-ray absorption fine structure, 159, 171
  infrared, 159, 171
  laser-induced fluorescence, 163
  laser spectroscopy cell sorter, 173
  low-energy electron diffraction, 158, 159, 171–172
  for micelle studies, 164
  for microstructure characterization, 158–159, 162, 171–172
  for monitoring deposition/etching processes, 163
  neutron spin-echo, 164
  nuclear magnetic resonance, 159, 171
  photon correlation, 164
  Raman, 159, 171
  small-angle neutron scattering, 124, 164
  solid-state nuclear magnetic resonance, 159
  structure-permeability probes, 168
  ultraviolet photoelectron, 158, 172
  x-ray photoelectron, 124, 158, 162, 168, 171, 172
Storage media, *see* Recording and storage media
Sulfur oxides from combustion processes, 117
Supercomputers
  applications, 139–140, 143
  artificial intelligence, 136–139
  availability, 137–138, 152
  Cray I class, 137
  expert systems, 136, 137
  hypercube architecture, 140–141
  need for, 173–174
  speed and capabilities, 136–138, 152
Superconductors
  ceramics in, 49, 56
  chemical manufacturing processes for, 49
  cooling, 49
  high-temperature, 2, 49, 52
  international competition in, 52
  materials, 49, 56
  metal oxide, 56
  uses, 49
  see also Semiconductors
Surface and interfacial engineering
  biological, 2, 17, 27–28, 155
  in catalytic and electrode reactions, 155, 159–160
  in ceramics, 166–167
  characterization techniques, 158–159
  in colloidal systems, 163–164
  in composites, 64, 72, 156

in concrete and cement, 167
  in film deposition, 58
  fluid, 163, 166
  in fuel cell technology, 161
  importance, 154–156
  lubricant interaction with, 156
  in microcircuit processing, 162
  multiple, structuring of, 155
  in natural resources recovery, 156
  properties and processes, 155
  research needs and opportunities, 6, 72, 156–174
  role in materials chemistry, 72
  solvent/polymer, 55
  in surfactants, 163
  tissue-implant, 27
Surfactants
  in cement and concrete, 167
  di-tail and tri-tail, 164
  in enhanced oil recovery, 154, 163
  monolayer-forming, 164
  multifunctional, 164
  property control measures, 165
  research opportunities, 163–164
  resource recovery applications, 83–84
  superplasticizers, 167
Synfuels
  catalytic conversion to liquid fuels, 87
  Fischer-Tropsch process, 87
  methanol to gasoline, 89–90, 93, 158–159
  natural gas to gasoline, 90
  production process, 86–88, 90–91
  uses, 25

T

Thin films
  controlled permeability, 154
  deposition processes, 2, 41, 50, 54, 56–58
  on interconnection devices, 2, 57
  low-temperature methods, 56
  mathematical modeling of processes, 58
  on optical fibers, 2, 57, 156
  pharmaceutical applications, 154
  on recording/storage media, 2, 46, 57, 70
  organic, 41
  property determinants, 58, 156
  research needs on, 58
  silicon dioxide, 41
  surfactant applications in processing of, 164
  U.S.-Japanese competition, 50
Tissue culture, *see* Cell/tissue culture
Tissue plasminogen activator, 21, 22
Tissues, *see* Artificial organs, tissues, and fluids; Prostheses
Training, *see* Education/training
Transmission electron microscopy, 159
Transportation, environmental impacts of fuel combustion for, 109–110

U

Ultrapurification
  for cell culture processes, 21
  for electronic, photonic, and recording materials and
      devices, 2, 41, 54
  of silicon for semiconductors, 41
Union Carbide, UNIPOL process, 63
United Kingdom
  biotechnology institutes, 25
Unocal Corporation, 96
U.S. competitiveness
  adhesives, 71
  biotechnology, 25
  ceramics, 71
  chemical processing industries, 11, 13, 38, 50
  composite manufacturing and processing technology, 71
  interconnection and packaging, 52
  liquid crystals, 71
  lubricants, 71
  microelectronics, 2, 48, 50
  optical technologies, 2, 45, 50
  photovoltaics, 2
  polymers, 70–71
  recording media, 2, 50–52
  superconductors, 52

Uranium-235
  scale-up of manufacturing process, 11

V

Vaccines, *see* Pharmaceuticals
Vesicles, 163

W

Waste management
  biological treatment, 17, 24, 122–123, 125
  multimedia approach, 5, 129
  nuclear waste repositories, 94
  regulation of, 110, 123
  site remediation, 124–125
  soil decontamination, 124–125
  Superfund Program, 111
  see also Hazardous wastes
Wastes
  metal/mineral recovery from, 98
  municipal solid, as an energy source, 91–92

Z

Zeolites, 156